Materials for Lightweight Constructions

This book presents the key concepts and methods involved in the development of a variety of materials for lightweight constructions, including metals, alloys, polymers, and composites. It provides case studies and examples to explain strategies adapted for specific applications of the materials and covers traditional to advanced manufacturing concepts of lightweight materials, including 3D printing. It also illustrates the fundamentals and usability of biodegradable materials for achieving a greener environment, as well as possibilities of green manufacturing.

- Covers the fundamentals of a range of materials used for lightweight constructions
- Discusses fabrication and testing of materials
- Addresses relevant concepts of 3D printing and biodegradable materials
- Explores analysis of the failure mechanism of materials used in various applications
- Identifies the applicability of materials to a variety of situations

Materials for Lightweight Constructions will suit researchers and graduate students in materials science, mechanical engineering, construction, and composites.

Materials for Lightweight Constructions

Edited by
S. Thirumalai Kumaran
Tae Jo Ko
S. Suresh Kumar
Temel Varol

CRC Press
Taylor & Francis Group
Boca Raton London New York

CRC Press is an imprint of the
Taylor & Francis Group, an **informa** business

First edition published 2023
by CRC Press
6000 Broken Sound Parkway NW, Suite 300, Boca Raton, FL
33487-2742

and by CRC Press
4 Park Square, Milton Park, Abingdon, Oxon, OX14 4RN

CRC Press is an imprint of Taylor & Francis Group, LLC

ISBN: 978-1-032-17173-9 (hbk)
ISBN: 978-1-032-17174-6 (pbk)
ISBN: 978-1-003-25210-8 (ebk)

DOI: 10.1201/9781003252108

Typeset in Sabon
by SPi Technologies India Pvt Ltd (Straive)

Contents

Preface

The use of materials in the various fields is inevitable. However, developing technologies have allowed for the use of lightweight materials for the desired applications. It is now becoming a part of the regular materials' vocabulary. Because of the unique nature of the materials, more advanced research has continued in an effort to reduce the weight of structures. There is increasing demand for such lightweight materials in a variety of sectors that minimize energy consumption without compromising material quality. The use of polymer nanocomposites, such as polymer/carbon nanomaterials, in the aerospace industry has grown in the last few years because of the strength of the components combined with their lightweight features.

Each chapter in this book has been extended with new information, which goes from basic to in-depth discussion and conclusion. This book intends to maintain a balance between conventional processing of materials and design, as well as the recent developments vis-à-vis achieving the immediate needs. This book is intended for undergraduate students, postgraduate students, and research scholars who wish to acquire a broad knowledge of lightweight materials. In addition, industry personnel will also glean ample information from this book, as it will help them comprehend and apply the concepts to practical solutions in better ways. The references cited in this book will also help them find additional information to go deeper into topics that are of interest to them.

S. Thirumalai Kumaran
Tae Jo Ko
S. Suresh Kumar
Temel Varol

Foreword

MATERIALS FOR LIGHTWEIGHT CONSTRUCTION

We stand today at a critical juncture. While the industrial revolution has given us a world of infinite possibilities, it has already created a world that is non-sustainable for the human race. Materials is at the heart of transformation and particularly the dawn of one civilisation after next. The ages of copper, bronze, silver, gold, and iron are behind us. Today we stand face to face with a diversity of materials – mostly synthesized using molecular chemistries. These new age materials have excellent functional properties – however, are they sustainable and green.

The core philosophy of going "green" is to reduce the impact of the material on the environment. It starts from the mining or synthesis of the starting blocks – be it a ore or a monomer. Then comes processing – either using furnaces driven by fossil rules or polymerisation using various toxic substances often leaving behind a difficult set of residues. Finally the usage of the material as a product and its end of life. Recycle-Reuse-Retain is the new age mantra that is driving sustainable manufacturing processes.

It is also noted that a new "sustainable" manufacturing process will be costly to start with. This is because the incumbent material enjoys a head start in terms of economies of scale, proven and sturdy supply chain and the established customer acceptance. It is imperative, therefore, to use modern tools such as 3D printing, Virtual reality and Artificial Intelligence to shorten the time to adoption of new materials and new processing technologies and make them commercially viable.

In light of the above, I am encouraged to go through the 11 chapters of the book. The topics are very pertinent to the needs of sustainable manufacturing. It is refreshing to see the take on automotive and construction and the attempt by the authors to connect the state of the art research with the needs of these sectors.

I am sure the readers will get a first-hand glimpse and will be better informed when making a choice of materials of the future as well as their manufacturing processes that are sustainable and green – right from the cradle to the next cradle !!

Sumitesh Das
Director, Research & Development

Tata Steel confidential information. Tata Steel does not accept any liability for damages that might arise as a result of use of the information in this document.

Editors

Dr. S. Thirumalai Kumaran received his M.Tech in Manufacturing Engineering (Gold medalist) from Government College of Technology, India, in 2008. He completed his Ph.D. degree in Mechanical Engineering from Kalasalingam Academy of Research and Education (KARE), India, in 2015. After completing his Ph.D., he served as Research Professor for one year at the School of Mechanical Engineering, Yeungnam University, South Korea. His credentials include three projects from UGC-DAE, DAE-BRNS and CVRDE worth of Rs.42,62,500/- and he has also published over 130 SCI/Scopus indexed journals with over 1350 citations. Recently, he was awarded "Young Scientist in Mechanical Engineering" and the "Dr. Sarvepalli Radhakrishnan Best Teacher Award" in appreciation of the dedication and commitment in teaching and research in Mechanical Engineering. Currently he is acting as Associate Professor in the Department of Mechanical Engineering, PSG iTech, Coimbatore.

Prof. Tae Jo Ko is a Professor in Mechanical Engineering at Yeungnam University, South Korea. He received his Ph.D. in Mechanical Engineering from POSTECH, South Korea. His research interests include micro-cutting process, non-traditional machining, surface texturing, bio-machining, hybrid EDM-milling process, deburring process of CFRP composite, and CFRP drilling. He recently launched new research into rechargeable batteries and digital twins for manufacturing. He is an Editor in the *International Journal of Precision Engineering and Manufacturing*—impact factor of 2.106 (2020). His research is recognized globally, and he has published over 200 articles in SCI/Scopus indexed journals. He has also received national and international project funding and traveled across the globe to deliver keynote lectures. Currently, he is acting as a Visiting Professor in Hunan University of Science and Technology, China.

Dr. S. Suresh Kumar completed his postgraduate degree in Manufacturing Engineering from Alagappa Chettiar College of Engineering and Technology, India, in 2012. He received his Ph.D. in Mechanical Engineering from Kalasalingam Academy of Research and Education (KARE), India, in 2016. He visited Karadeniz Technical University (KTÜ), Turkey, for his post-doctoral experience in 2018. He has 11 years of teaching experience and 12 years of research experience. His areas of research include fabrication, processing, and characterization of materials. He has published in more than 40 international journals and presented in conferences. At present, he is Associate Professor in the Department of Mechanical Engineering in KARE.

Dr. Temel Varol obtained his Ph.D. in Mechanical Engineering from the Karadeniz Technical University (KTÜ), Turkey, in January 2016, where he began education and research studies in the Department of Metallurgical and Materials Engineering. He is Associate Professor in the Department of Metallurgical and Materials Engineering and also the Vice Dean of Engineering Faculty at KTÜ. His current research interests include powder metallurgy, metal-based additive manufacturing, mechanical alloying, coating, non-ferrous metals, metal matrix composites, electrical materials, porous structures, biomaterials, and conductive materials. His research outcomes are recognized both nationally and internationally, as is evident from his more than 70 articles, many of which have been published in high-impact journals and well cited. Dr. Varol also has an international patent on the development of new type electrical contact materials. He has expertise in national and international funding and has established numerous fruitful national and international collaborations. Dr. Varol is a member of the Chamber of Mechanical Engineers in Turkey.

Contributors

P. Amuthakkannan
Department of Mechanical
  Engineering
PSR Engineering College of
  Engineering
Tamil Nadu, India

K. Anand Babu
Department of Production
  Engineering
National Institute of Technology
Tiruchirappalli, Tamil Nadu, India

V. Arumugaprabu
Department of Mechanical
  Engineering
Kalasalingam Academy of Research
  and Education
Krishnankovil, Tamil Nadu, India

M. Aruna
College of Engineering and
  Computing
Al Ghurair University
Dubai, UAE

K. Arunprasath
Department of Mechanical
  Engineering
PSN College of Engineering and
  Technology
Tirunelveli, Tamil Nadu, India

S. K. Ashok
Department of Automobile
  Engineering
Dr. Mahalingam College of
  Engineering and Technology
Pollachi, Tamil Nadu, India

N. B. Karthik Babu
Department of Mechanical
  Engineering
Assam Energy Institute, A Center
  of Rajiv Gandhi Institute
  of Petroleum Technology
Sivasagar, Assam

M. Bhuvanesh Kumar
Department of Production
  Engineering
NIT, Tiruchirappalli, Tamil Nadu,
  India
Department of Mechanical
  Engineering
Kongu Engineering College
Erode, Tamil Nadu, India

Oisik Das
Structural and Fire Engineering
  Division, Department of Civil,
  Environmental and Natural
  Resources Engineering
Luleå University of Technology
Sweden

Milon Selvam Dennison
Department of Mechanical
    Engineering
Kampala International University
Western Campus, Uganda

K. Jayakrishna
School of Mechanical Engineering
Vellore Institute of Technology
Vellore, Tamil Nadu, India

Joshua Jeffrey
School of Civil Engineering
Vellore Institute of Technology
Vellore, Tamil Nadu, India

Akesh B. Kakarla
School of Computing, Engineering
    and Mathematical Sciences
La Trobe University
Bendigo, Australia

Ing Kong
School of Computing, Engineering
    and Mathematical Sciences
La Trobe University
Bendigo, Australia

V. Lakshmi Narayanan
School of Mechanical Engineering
Vellore Institute of Technology
Vellore, Tamil Nadu, India

V. Manikandan
Department of Mechanical
    Engineering
PSN College of Engineering and
    Technology
Tirunelveli, Tamil Nadu, India

Mohammad Mehdi Hosseini
Faculty of Mechanical Engineering-
    Energy Division
K. N. Toosi University of Technology
Tehran, Iran

Rhoda Afriyie Mensah
Structural and Fire Engineering
    Division, Department of Civil,
    Environmental and Natural
    Resources Engineering
Luleå University of Technology
Sweden

M. Mohanraj
Department of Mechanical
    Engineering
Hindusthan College of Engineering
    and Technology
Coimbatore, Tamil Nadu, India

M. Natesh
Department of Mechanical
    Engineering
V.S.B. Engineering College
Karur, Tamil Nadu, India

Satya G. Nukala
School of Computing
Engineering and Mathematical
    Sciences
La Trobe University
Bendigo, Australia

Hitesh Panchal
Mechanical Engineering
    Department
Government Engineering College
Gujarat, India

Yash Panchal
Smart Manufacturing Lab,
    Mechanical Engineering Discipline
IIITDM, Jabalpur, Madhya Pradesh,
    India

K. Ponappa
Smart Manufacturing Lab,
    Mechanical Engineering Discipline
IIITDM, Jabalpur, Madhya Pradesh,
    India

A. Pugazhenthi
Department of Mechanical
  Engineering
University College of Engineering,
  Dindigul
Tamil Nadu, India

C. Pradeep Raja
Department of Mechanical
  Engineering
National Institute of Technology
Tiruchirappalli, Tamil Nadu,
  India

T. Ramkumar
Department of Mechanical
  Engineering
Dr. Mahalingam College of
  Engineering and Technology
Pollachi, Tamil Nadu, India

Abhimanyu Singh Rathor
School of Civil Engineering
Vellore Institute of Technology
Vellore, Tamil Nadu, India

J. Ronald Aseer
Department of Mechanical
  Engineering
National Institute of Technology
Puducherry, India

P. Sathiya
Department of Production
  Engineering
National Institute of Technology
Tiruchirappalli, Tamil Nadu,
  India

D. Satish Kumar
Department of Mechanical
  Engineering
Saveetha Institute of Medical and
  Technical Sciences
Chennai, Tamil Nadu, India

M. Selvakumar
Department of Automobile
  Engineering
Dr. Mahalingam College of
  Engineering and Technology
Pollachi, Tamil Nadu, India

Vigneshwaran Shanmugam
Department of Mechanical
  Engineering
Saveetha Institute of Medical and
  Technical Sciences
Chennai, Tamil Nadu, India

A. Sofi
Department of Structural and
  Geotechnical Engineering
School of Civil Engineering, Vellore
  Institute of Technology
Vellore, Tamil Nadu, India

Ali Sohani
Faculty of Mechanical Engineering-
  Energy Division
K. N. Toosi University of Technology
Tehran, Iran

A. Soundhar
Department of Mechanical
  Engineering
Indian Institute of Technology
  Guwahati
Assam, India

Pon Janani Sugumaran
Department of Biomedical
  Engineering
National University of Singapore
Singapore

R. Sundarakannan
Institute of Agricultural Engineering
Saveetha School of Engineering,
  SIMATS
Chennai, Tamil Nadu, India

Burçin Atilgan Türkmen
Department of Chemical
    Engineering
Bilecik Şeyh Edebali University
Bilecik, Turkey

G. Velmurugan
Institute of Agricultural
    Engineering
Saveetha School of Engineering,
    SIMATS
Chennai, Tamil Nadu, India

M. Vijayakumar
Department of Mechanical
    Engineering
PSN College of Engineering and
    Technology
Tirunelveli, Tamil Nadu, India

Jaroslav Vrchota
Department of Management,
    Faculty of Economics
University of South Bohemia in
    Ceske Budejovice
Czech Republic

Chapter 1

Introduction to lightweight materials

Vigneshwaran Shanmugam
Saveetha Institute of Medical and Technical Sciences, Chennai, India

Rhoda Afriyie Mensah
Luleå University of Technology, Luleå, Sweden

N. B. Karthik Babu
Assam Energy Institute, A Center of Rajiv Gandhi Institute of Petroleum Technology, Sivasagar, Assam

C. Pradeep Raja
National Institute of Technology, Tiruchirappalli, India

J. Ronald Aseer
National Institute of Technology, Puducherry, India

A. Pugazhenthi
University College of Engineering Dindigul, Dindigul, India

D. Satish Kumar
Saveetha Institute of Medical and Technical Sciences, Chennai, India

Oisik Das
Luleå University of Technology, Luleå, Sweden

CONTENTS

DOI: 10.1201/9781003252108-1

1

1.1 OVERVIEW OF MATERIALS

The evolution of materials determines the evolution of the human stage. It should be noted that human evolution has been classified according to material as Stone Age, Bronze Age, Iron Age, and Steel Age. The evolution of materials science is an important area for the development of various industrial sectors. The evolution of materials from clay-based ceramics to modern advanced materials depicts the evolution of human civilization. As a result, the material system is intimately linked to human development. In modern applications, advanced engineering materials are utilized, which deliver enhanced performance. The materials have been used in aircraft, construction, space, and military applications. Metals, semiconductors, ceramics, and polymers are the four main material classifications. Aside from these material systems, notable advanced materials such as nanomaterials, biomaterials, and energy materials can also be found. All of these materials differ in terms of structure, processing method, and properties. Lightweight materials play a dominant role in aerospace and automotive applications because they are lighter and tougher, resulting in improved performance. In these applications, lightweight materials such as composites have been widely used. For example, a commercial plane, such as the Dreamliner, is made up of 50% composite materials. The manufacturing of these materials is also less expensive, but it is a labor-intensive process. Lightweight aluminum alloys in automobile applications reduce vehicle weight, save fuel, and are more temperature-resistant than conventional steel materials. These advancements were made with existing technology, so it's no surprise that in the future, the scope of these materials can be found in a wide range of applications.

1.2 IMPORTANCE OF LIGHTWEIGHT MATERIALS

One major problem with conventional materials (such as steel, cast iron, and so on) is weight. This single parameter leads to increase in cost, fuel consumption (in case of automotive), risk during assembly, etc. In many circumstances, it's possible to replace conventional material with low-weight materials such as aluminum alloy, magnesium alloy, fiber-reinforced polymer composites, and so on. Such possibilities exist in different sectors like automotive, biomedical, construction, and aerospace applications. The following sections cover the importance of lightweight materials in the above-mentioned applications.

The specific properties (property/density) such as specific strength and specific modulus of high-performance fiber-reinforced polymer composites, mainly carbon-reinforced materials, are greater compared to conventional aerospace metallic alloys. As a result, larger weight savings can be achieved, which leads to obtain better performance, superior payloads, and fuel savings.

In addition, polymer composite is well known for its non-corrosiveness and extraordinary fatigue resistance. Though aluminum alloy is light in weight, its corrosion is a major cost, and scheduled maintenance is required for both commercial and military aircraft. The corrosion resistance of polymer composites can result in main savings in supportability costs. Furthermore, assembly costs can account for up to 50% of the cost of an airframe. Drastic reduction of assembly labor and fasteners used in structural application is possible with low-weight composite materials.

Automotives can be made lighter by cutting down heavy parts and replacing them with a low-weight material without functional degradation. It is possible to reduce overall fuel consumption when heavy metal parts can be replaced by advanced lightweight materials. According to the Department of Energy, reducing a vehicle's weight by 10% can improve fuel economy by 6% to 8%. "There are literally hundreds of different technologies that can be brought to bear to help improve fuel efficiency; however, all of them begin with lightweighting," said a spokesman with the aluminum company, Alcoa Inc. [1].

1.3 LIGHTWEIGHT MATERIALS IN THE TWENTY-FIRST CENTURY

Several novel lightweight materials with varying physical, thermal, and chemical properties have been developed since the beginning of this century [2]. This development has paved the way for engineers to have access to a wide range of materials suitable for various applications such as wind, marine, automotive, etc. In particular, lightweight materials are mostly applied in the automotive and aerospace industries where lighter parts are analogous to fuel efficiency and overall performance. Lightweight materials have high strength-to-density ratio and are required for the reduction of the weight of a part without compromising strength and efficacy. Evans et al. [3] listed four parameters that influence weight reduction in product manufacturing. According to the research, material selection, shape utilization, topology optimization, and multi-functionality greatly affect the achievement of minimal weight. Among the four factors, material selection is paramount. This is undoubtedly reflected in the drastic increase in the global demand for lightweight materials in recent years. It is also expected to rise at a tremendous rate in the future following the production and continual usage of "smart" products. According to market research on lightweight materials, the global economic value for the year 2022 is estimated at US$189,076 million, thus a CAGR of 5.8% from US$128,697 million in 2015 [4].

Due to the increase in greenhouse gas (GHG) emissions and the depletion of fossil fuels, materials that impart sustainability are being sought after. In this regard, several researchers have studied the amount of GHG emissions released from lightweight materials in recent years. In the research of Kawajiri et al. [5], the authors noted that light weighting, especially in vehicles, does not necessarily correspond to light weight emissions. For instance, the GHG coefficient of Al is 10.26 kg-CO_2 eq/kg, while that of regular steel is 1.77 kg-CO_2 eq/kg [6]. Hence, the replacement of steel with Al will definitely increase GHG emissions. In their analysis, it was observed that high strength steel and carbon fiber-reinforced polymer (CFRP) are effective alternatives compared to magnesium alloys and regular steel even though Mg and its alloys are the lightest [5, 7].

Lightweight materials are classified under metal alloys (Aluminum [Al], high strength steel, magnesium [Mg], titanium [Ti], beryllium-based alloys), composites (polymer-matrix, ceramic-matrix, metal-matrix), and polymers (polycarbonate, polypropylene, etc.) [8].

1.3.1 Metal alloys

Metal alloys are relatively cheap depending on the type and have several desirable qualities that increase their applicability. They have high electrical and thermal conductivities, high stiffness, resistance to corrosion and oxidation, and significantly low density [9]. The density of metal alloys ranges from 1.5 g/cm³ to 4.5 g/cm³ with titanium alloys possessing the highest density. Metal alloys are selected based on their end use. For structural applications, Mg alloys are preferred to Al and Ti alloys, while steel and polymers are commonly used in the automotive industry. Similarly, the aerospace sector uses Mg, Ti, and Al alloys [8]. Technological advancement in material research has brought about modifications in conventional metal alloys and the development of novel ones. For instance, aluminum is the most commonly used additive in magnesium alloys; however, other techniques such as the inclusion of nanoparticles are being exploited [10]. Some of these materials and their properties are outlined in this section.

Titanium alloys have a high strength-to-weight ratio, good biocompatibility, corrosion resistance, and high operating temperatures that make them suitable for elevated temperatures. It is applied in the biomedical, paint, and aerospace industries, etc. [11, 12]. The critical issue encountered in the usage of titanium alloys is the high production cost [13]. For over a decade, research on titanium has focused on reducing the cost and providing alternative production routes for obtaining titanium alloys [14]. Furthermore, techniques for manufacturing high-quality titanium products such as plastic forming technologies and simulations of the behavior of the alloys have been explored [15]. The elements added to titanium to produce alloys are classified under alpha and beta stabilizers. Examples of the former are Al, O, N, C and the latter, V, Mo, Ni, Fe, Cr, etc. [16]. However, for low-cost titanium alloys, iron is preferred owing to the high strength, diffusivity, and enhanced sintering capability [17].

Aluminum alloys are the most widely used replacement for steel. These alloys are responsive to a broad range of production processes, hence they are available in many forms: castings, extrusions, stampings, forgings, impacts, and machined components [18]. They are non-magnetic and non-sparking. The density (2.69 g/cm^3) and elastic modulus (69GPa) are one third that of steel [9]. The mechanical properties of the alloys depend on the composition, temper, solidification rate, and the shape of the part. One key advantage of the utilization of aluminum alloys is the low energy required for recycling. In addition, recycled aluminum significantly minimizes greenhouse gas emissions [19].

Beryllium alloys are used in critical applications where high precision is required, especially in the military and aerospace industries. They are used for making optical and precision instruments. Beryllium has high heat capacity, specific modulus, thermal conductivity, and yield strength that makes it suitable for these applications [8]. Beryllium alloys such as AlBeMet AM 162 have lower densities and are about four times stiffer compared to aluminum and titanium [20].

1.3.2 Composites

Composites consist of at least two homogenous materials: a matrix (polymer, metal, or ceramic) and reinforcement (fiber or particulate). These composites combine the properties of the individual constituents to produce significant performance benefits. However, composites are not moisture-resistant and degrade when exposed to high temperatures due to the high amount of raw material. Additionally, poor interfacial bonding between the matrix and the reinforcement will minimize the strength of the composite. Some examples of lightweight composite materials are aluminum-silicon-carbide, beryllium/beryllium oxide, and aluminum beryllium composites. Metal composites have high corrosion and fatigue resistance, which makes them suitable for many engineering applications.

Previous studies have shown that the addition of small amounts of nano-sized reinforcements to metals such as magnesium enhances the wear resistance, damping properties, and mechanical properties including strength and ductility [21–24]. The main challenge encountered in the fabrication of these composites is the low wettability exhibited during synthesis [24]. Gupta et al. [7] investigated the effect of ceramic (Mg/Al$_2$O$_3$, MgSiC, MgZnO, etc.), metallic (MgCu), and hybrid (e.g., MgZr$_2$ + Cu) reinforcements on the mechanical properties of pure Mg. The composites were synthesized by disintegrated melt deposition and hybrid microwave sintering. The results showed that the addition of nano particles increased the hardness of Mg. In addition, the metallic reinforcements successfully increased the strength; however, low ductility was observed. This was attributed to the formation of clusters of reinforcements that acted as stress concentration points. Overall, the hybrid samples, Mg + ceramic + metallic nano composites, produced the optimal results. Thus, the combination exhibited an increased strength and simultaneously improved the ductility of Mg.

Carbon fiber, graphite fiber, and Kevlar fiber-reinforced polymers are some of the widely used lightweight polymer-matrix composites. These fiber-reinforced polymers have a high strength-to-weight ratio, stiffness, and are resistant to chemicals. Among other applications, they can be used for fabricating or repairing lightweight concrete members with enhanced strength and service life compared to the conventional ones [25–27]. Hybrid fiber polymer composites with multiple reinforcements in single or multiple matrices prepared by injection molding enhances the mechanical properties, thermal properties, water absorption rate, and wear rate of single fiber-reinforced composites [28]. Banana/sisal-reinforced epoxy composites, palmyra palm leak stalk/jute fiber-reinforced polyester composite, jute/oil palm epoxy composite, etc., are some examples of the hybrid composites.

A novel organic geopolymer-based hybrid foam was fabricated by Roviello et al. [29, 30] by combining metakaolin and an alkali silicate solution with mixtures of dialkylsiloxane oligomers (GHF-MK). The foaming agent used for the reaction was Si^0 powder at varying weight percentages. The resultant foams from the experiment were lightweight and had densities ranging from 0.25 to 0.85 g/cm^3, which is significantly lower than that of metal alloys. It was seen that the foams displayed excellent mechanical properties, thermal conductivity, and fire resistance.

A major drawback of the utilization of polymer-based composites is the low thermal stability of the polymers, which limits the application, especially upon exposure to elevated temperatures [31].

1.3.3 Polymers

Polymers are made up of repeating units of monomers. They are grouped into thermoplastics, thermosets, and elastomers. Lightweight polymers have high toughness, durability, formability, impact resistance, and release fewer greenhouse gases. Materials such as polyester, vinyl ester, polystyrene, epoxy, and polyurethane are some of the lightweight polymers [32]. Polymers for lightweight applications are further classified into high-performance polymers, polymers for weight reduction, polymer sandwich panels, and polymer/metal hybrid systems. A diverse range of polymeric materials can be employed for various applications, and the manufacturing techniques required for their production vary greatly. During manufacturing, shaping processes must be chosen according to the materials being used and the product design [33]. It has been argued that the replacement of conventional materials such as steel with lightweight plastics could reduce the overdependence on fossil fuels [34]. However, the low modulus and strength limit the applicability of polymers. In addition, their mechanical properties such as ductility, etc., are influenced by temperature and/or humidity.

1.4 OPPORTUNITIES FOR NEW MATERIALS AND RESEARCH

The thirst for new materials for various applications has persisted throughout the evolution of human civilization. The exploration and development of novel materials is an ongoing process in nature as the demand for novel materials grows. At present, the demand for novel materials stems from a variety of fields, including tools and equipment used in human survival, structural applications, armor materials for military applications, biocompatible materials for organ transplants, biomedical instrumentation, high strength-to-weight ratio materials in aerospace industries, energy development and storage applications, and so on. The ongoing search for and development of new materials is fueled by two factors: curiosity and demand in specific applications. To name a few, recent research on novel materials includes composites (metal, polymer, or ceramic) with varying reinforcement in different forms/ratios, nano materials (carbon nano tubes, graphene, etc.), high strength alloys, biodegradable/biocompatible materials, materials for hydrogen storage, raw materials for developing complex shapes using 3D printing, and so on.

The availability of a wide range of materials with varying properties and manufacturing processes creates room for the development of new lightweight materials. With the increase in global warming, sustainable lightweight materials with excellent properties that could capture and sink carbon dioxide is a necessity. Biochar, a biomass-derived product that has gained popularity over the years, could be used as a reinforcement in lightweight composites [35]. Biochar produced from high-temperature pyrolysis is fire-resistant, has a large specific surface area, low bulk density, and excellent physical and mechanical properties [36]. The replacement of a fraction of a material with the low-dense biochar allows for the production of lighter materials with enhanced properties. Hence, biochar has been included in the production of concrete structures, polymer-based composites, etc. [37, 38]. While the technology for biochar-based products is well established, significant market entry barriers exist due to the lack of education on the potential of biochar [38].

Another material recently gaining ground is graphene obtained from the exfoliation of graphite. Graphene is one of the thinnest and lightest materials produced and known as a wonder material due to its exceptional properties [39]. Integration of graphene into other matrices to form composites has gained popularity because only a small weight percentage (<1%) is needed to bring about added benefits and improved performance. Although graphene is expensive, there are companies including TLC Graphene, etc., which integrate graphene into plastic resin composites while still making them commercially feasible from a financial standpoint. Other forms of graphene such as aerographene, aerographite, 3D graphene, and novel lightweight materials like carbyne and metallic microlattice have been fabricated over the years. Extensive research on the machining processes and characterization of these materials could create an avenue for producing lightweight materials with desirable properties.

1.5 INDUSTRY REVOLUTION TOWARDS MATERIALS AND STRUCTURES

The industrial revolution, which began in the United Kingdom in the eighteenth century, was the driving force behind development in the material system. The first Industrial Revolution, which lasted from 1760 to 1840, was a period of significant technological, cultural, and socioeconomic transformation. Rural societies became more industrialized during this period. Notable improvements in the material system were noted after the Second Industrial Revolution happened, roughly from 1870–1914. Steel was an expensive material prior to the industrial revolution; the Bessemer converter was then invented, allowing steel to be mass produced at a lower cost. It is safe to say that iron and steel effectively fueled the industrial revolution.

The impact of the industrial revolution resulted in the advancement of the materials system, paving the way for the arrival of different materials such as cast iron, steel, and glass, with which architects and engineers rearranged the concept of many applications. Europe, and particularly, Britain, was the pivot of this revolution, owing to the abundant supply of coal and iron ore. Steel became popular because it is lighter, less expensive, and stronger than iron, making it suitable for a wide range of applications, mainly in railways. The use of metallurgical products during times of war was also noteworthy.

It should be noted that the development of large-scale iron and steel production, as well as the widespread use of machinery in manufacturing, occurred during the Second Industrial Revolution. Steel-based railroads were paved during this time. The Second Industrial Revolution changed the perspective of engineering applications by replacing iron utilization in a great number of applications.

1.6 APPLICATIONS

Lightweight materials are adopted in a myriad of applications. Potential applications include aeronautical, marine, automotive, construction, food packaging, medicine, etc. Nevertheless, recent material development and research have introduced more opportunities for product design. For instance, the production of lighter smart phones, computers, batteries, and other electronic gadgets requires the use of high-quality materials that are energy-efficient, eco-friendly, and can help attain weight-reduction goals. In addition, the industries should be equipped with the manufacturing systems needed to produce lightweight parts to ensure an easy transition.

Materials such as metals, polymers, ceramics, plastics, and various combinations of metallic alloys and biomaterials are now widely utilized in biomedical applications. The application of such materials for treatment of underlying disease, physical injury, or implant has been practiced for decades, but its true potential has only recently been realized, and advanced technology such as 3D printing, digital imaging, tissue engineering, and implants has helped us combine the various benefits of the biomaterials for medical treatments. An important application such as in the areas of the cardiovascular system, dental implants, artificial cartilage, artificial ligaments, and joint prosthetics is a notable development in the implementation of lightweight materials in biomedical application.

The pacemaker and stent are important in cardiovascular-based applications as they are used to treat cardio-related diseases. A pacemaker is a lightweight electrochemical system that monitors and maintains the circulation and pumping of blood in the heart. The components of pacemakers have generally been made from platinum alloys, but to reduce the skin irritation and weight of the components of the pacemaker, the platinum alloys have been replaced with more corrosion-resistant composites.

A stent is a device inserted into a vessel or duct to prevent constrictions due to disease or injury or blockage. A stent can be used as a drug delivery system or to treat a cardio system block (thrombosis) where a blood clot blocks the flow of blood; a stent can make a passage for recirculation of blood or remove the clot. Stents are also used during angioplasty and even during surgery to keep the operating channels open. Most stents are made of stainless steel and alloys of iron, titanium, magnesium alloys, and plastics. The latest advancements have provided lightweight polymer matrix composites that replace the materials previously used for stents because of their significant flexibility in design, low cost, and biodegradability. These modern polymer matrix composite stents provide good short-term structural integrity, and they are readily absorbed by surrounding tissues. With the knowledge of human systems and the reactions to foreign materials in contact with human tissues, we can also prevent various types of irritations that occur during contact with human tissues and stents [40].

Dental injuries due to accidents and cavity problems are treated using tooth filling and dental implants. Tooth cavities are filled using silver-mercury amalgam that is toxic, and acrylic-based resins that are used as artificial teeth possess low mechanical strength. And hence, to replace these materials, polymer matrix composites are currently used for implants.

Polymer matrix composites are also more compatible than traditionally used biomaterials. Composites such as poly (2-hydroxy ethyl methacrylate) [PHEMA] reinforced with polyethylene terephthalate (PET) synthetic fibers are used as a replacement for cartilage damaged due to physical accidents, birth defects, and disease. Cartilages are soft bone structures whose main function is to connect bones, and they are highly flexible during childhood and develop into bone structures during adulthood. They can become damaged due to accidents, repeated cyclic loading, or sometimes because of birth defects.

By varying the compositions of the biomaterials, artificial cartilage that matches the properties of the original human cartilage can be produced and implanted. But indeed, the artificial cartilage has its drawbacks, as it is man-made and does not last as long as human cartilage. Artificial cartilages fail due to wear and cyclic loadings that can be prevented or prolonged by injecting hydrogels (polymeric biomaterials) that reduce wear and promote healing [41].

Similarly, ligaments damaged due to injury because of cyclic loading can also be replaced using artificial ligaments. Damage to ligaments is common among athletes, cyclists, and workers that lift heavy weights. A biodegradable composite made of polylactic acid and hyaluronic acid ester is used as a replacement in this case of an artificial ligament. Combinations of metals, polymers, composites, and ceramics are used as joint prosthetics, such as with hip joints in which the stem is made of metal alloy (good strength) and the ball a combination of composite material (high wear resistance). Hence, we can clearly see that these biocomposite materials, metallic alloys, and ceramics can be tailored to achieve the desired properties and serve the specific biomedical application.

With recent developments in technology, we can now identify the special type of materials such as smart polymers that are responsive to external stimuli. These smart polymers respond to the external surrounding environment and change their property. The stimuli could be due to changes in temperature, magnetic field, electric field, pH, or light field. But whatever the stimuli are, they can alter their chemical and physical property instantly and hence are mostly used as an ON/OFF in tissue engineering, gene therapy, and diagnostics.

While discussing materials widely used in the biomedical field, we can see that titanium and its alloys are utilized due to their light weight, corrosion-resistant nature, and high tensile strength. When titanium is combined with Nickel (Ni), we get Nitinol, a smart material that has shape memory characteristics and can be used in dental implant wirings. Tantalum and various other amorphous alloys are also used in biomedical engineering, and they make excellent stents for good x-ray visibility and amorphous alloys for better strength and corrosion resistance. Tissue engineering combined with 4D printing enables us to regenerate and replace the fully damaged or partially damaged tissue. Several aesthetic treatments such as facial reconstruction and disfiguration can also be treated with a combination of tissue engineering and bioprinting. Magnesium is a lightweight material widely used in the aerospace and automotive industries. Because of its lightweight characteristics, biocompatibility, and biodegradable nature, it is considered for bio-implants. With the combination of Zinc (biocompatible) and calcium, it can be an attractive material for medical applications [42].

Another most important material is the Polyether Ether Ketone (PEEK) that accomplishes the long-term demand of an ideal multipurpose material for biomedical applications. At present, PEEK is valued among the most important engineering polymers, all due to its outstanding properties such as high mechanical strength, better thermal stability, higher chemical resistance, good wear resistance, and anti-corrosive nature. PEEK also possesses the most desirable characteristic as the material for the future: its resistance to degradation. These properties are extremely desirable for applications in the biomedical industry. As far as the biomedical field is considered, PEEK has been progressively used as a biomaterial for orthopedic implants and prostheses since 1987 [43, 44]. PEEK is safe and does not encourage any mutagenic or cytotoxic activity as discovered from Vitro biocompatibility tests. PEEK does not produce any harmful reaction or release any harmful constituents to the human tissues, thus making it a bioinert material.

With all these positive aspects of PEEK, the scope of improvement is still relatively low, which owes to the hydrophobic nature of PEEK and limits the cell adhesion and absorption of protein on the surface, which ultimately reduces the wound healing capacity on osseointegration—the only issue that needs to be examined while considering PEEK's drawback. While combining PEEK with bioactive particles such as hydroxyapatite, we can make some modifications that will arouse cell attachment and proliferation [45]. Retrospective records of patients who underwent PEEK and titanium cranioplasty to compare the complication and failure rates between two types of implants were studied by Thien et al. [46], and they found the failure rate of PEEK cranioplasty (12.5%) was half that of Titanium (25%). The mechanical strength of PEEK for biomedical applications can be enhanced by reinforcing it with a proper filler. For example, carbon fiber-reinforced PEEK has Young's modulus in the range of the cortical bone (i.e., around 18 GPa), while the neat PEEK have Young's modulus of about 3–4 GPa. PEEK has excellent durability and increases in strength when it is combined with carbon nanofiber [47].

Lastly, the most abundantly available polymer with good potential is the straight-chain polysaccharide cellulose. Cellulose has good mechanical properties and its cost is low, but it is not widely used. It can produce numerous eco-friendly materials when combined productively. Nano cellulose has been used in the dressing of wounds due to its anti-infection properties but when combined with zinc it can be used to treat burn wounds. The wound healing nature of Bacterial cellulose (BC) and the antimicrobial behavior of zinc yields an efficient burn wound dressing. BC and zinc oxide nanocomposite-treated animals showed a significant 66% healing activity [48].

So many materials are available in the field of bio materials to treat various diseases and irregularities. Light weight, high mechanical strength, anti-corrosiveness, and biodegradability decide the role of the material in various applications. The opportunities for lightweight material in the biomedical field is high, which can be filled with the advancements of additive manufacturing, digital imaging, biocomposites, and smart materials.

REFERENCES

1. To Boost Gas Mileage, *Automakers Explore Lighter Cars - Scientific American*, (n.d.). https://www.scientificamerican.com/article/to-boost-gas-mileage-automakers-explore-lighter-cars/ (accessed January 21, 2022).
2. E. Karana, V. Rognoli, O. Pedgley, *Materials Experience Fundamentals of Materials and Design*, Elsevier Science Publishers B.V., Oxford, 2014.
3. A. G. Evans, Lightweight materials and structures, *MRS Bulletin* 26 (2001) 790–797. doi:10.1557/mrs2001.206.
4. S. Satsangi, Light weight materials market research, (2016) 110.
5. K. Kawajiri, M. Kobayashi, K. Sakamoto, Lightweight materials equal lightweight greenhouse gas emissions?: A historical analysis of greenhouse gases of vehicle material substitution, *Journal of Cleaner Production* 253 (2020) 119805. doi:10.1016/j.jclepro.2019.119805.
6. IDEA_v2.1.3 database, (2017).
7. M. Gupta, W. L. E. Wong, Magnesium-based nanocomposites: Lightweight materials of the future, *Materials Characterization* 105 (2015) 30–46. doi:10.1016/j.matchar.2015.04.015.
8. F. C. Campbell, *Lightweight Materials: Understanding the Basics, Materials Park*, ASM International, Ohio, 2012, n.d.
9. J. C. Benedyk, Aluminum alloys for lightweight automotive structures, in: *Materials, Design and Manufacturing for Lightweight Vehicles*, Elsevier, 2010: pp. 79–113. doi:10.1533/9781845697822.1.79.
10. A. I. Taub, A. A. Luo, Advanced lightweight materials and manufacturing processes for automotive applications, *MRS Bulletin* 40 (2015) 1045–1054. doi:10.1557/mrs.2015.268.
11. F. H. Froes, H. Friedrich, J. Kiese, D. Bergoint, Titanium in the family automobile: The cost challenge, *JOM* 56 (2004) 40–44. doi:10.1007/s11837-004-0144-0.
12. Y. Lu, L. H. Pan, G. S. Chen, S. L. Zhang, Strength of concrete with different contents of fly ash, *Journal of Shenyang University Technology* 31 (2009) 107–111.
13. C. Leyens, M. Peters, eds., *Titanium and Titanium Alloys*, Wiley, 2003. doi:10.1002/3527602119.
14. F. H. S. Froes, M. N. Gungor, M. Ashraf Imam, Cost-affordable titanium: The component fabrication perspective, *JOM* 59 (2007) 28–31. doi:10.1007/s11837-007-0074-8.
15. H. Yang, X. Fan, Z. Sun, L. Guo, M. Zhan, Recent developments in plastic forming technology of titanium alloys, *Science China Technological Sciences* 54 (2011) 490–501. doi:10.1007/s11431-010-4206-y.
16. M. O. Bodunrin, L. H. Chown, J. A. Omotoyinbo, Development of low-cost titanium alloys: A chronicle of challenges and opportunities, *Materials Today: Proceedings* 38 (2021) 564–569. doi:10.1016/j.matpr.2020.02.978.

17. L. Bolzoni, E. M. Ruiz-Navas, E. Neubauer, E. Gordo, Inductive hot-pressing of titanium and titanium alloy powders, *Materials Chemistry and Physics* 131 (2012) 672–679. doi:10.1016/j.matchemphys.2011.10.034.

18. F.-S. Hung, Design of lightweight aluminum alloy building materials for corrosion and wear resistance, *Emerging Materials Research* 9 (2020) 750–757. doi:10.1680/jemmr.19.00177.

19. A. Bull, Going Green: The Aluminum Perspective, in: *The American Metal Market's 3rd Annual Automotive Metals Conference*, Detroit, 2008.

20. T. B. Parsonage, Development of aluminum beryllium for structural applications, in: 1997: p. 102890E. doi:10.1117/12.279808.

21. M. Alderman, M. V. Manuel, N. Hort, N. R. Neelameggham, eds., *Magnesium Technology*, John Wiley & Sons, Inc., Hoboken, NJ, 2014. doi:10.1002/9781118888179.

22. S. C. Tjong, Novel nanoparticle-reinforced metal matrix composites with enhanced mechanical properties, *Advanced Engineering Materials* 9 (2007) 639–652. doi:10.1002/adem.200700106.

23. G. D. Cole, Magnesium, chemical & engineering news archive. *American Chemical Society* 81 (2003) 52. doi:10.1021/cen-v081n036.p052.

24. R. Casati, M. Vedani, Metal matrix composites reinforced by nano-particles: A review, *Metals* 4 (2014) 65–83. doi:10.3390/met4010065.

25. M. J. Shannag, N. M. Al-Akhras, S. F. Mahdawi, Flexure strengthening of lightweight reinforced concrete beams using carbon fibre-reinforced polymers, *Structure and Infrastructure Engineering* 10 (2014) 604–613. doi:10.1080/15732479.2012.757790.

26. Y. Liu, W. Liu, Y. Tong, T. Meng, Y. Su, Mechanism and preventive measures of external fire spread in Southwest Chinese traditional villages, *IOP Conference Series: Earth and Environmental Science*. 643 (2021). doi:10.1088/1755-1315/643/1/012149.

27. N. Pannirselvam, V. Nagaradjane, K. Chandramouli, Strength behaviour of fibre reinforced polymer strengthened beam, *Journal of Engineering and Applied Sciences* 4 (2009) 34–39.

28. M. K. Gupta, R. K. Srivastava, A review on characterization of hybrid fibre reinforced polymer composite, *American Journal of Polymer Science & Engineering* 4 (2016) 1–7.

29. G. Roviello, L. Ricciotti, O. Tarallo, C. Ferone, F. Colangelo, V. Roviello, R. Cioffi, Innovative fly ash geopolymer-epoxy composites: Preparation, *Microstructure and Mechanical Properties, Materials* 9 (2016) 461. doi:10.3390/ma9060461.

30. G. Roviello, C. Menna, O. Tarallo, L. Ricciotti, F. Messina, C. Ferone, D. Asprone, R. Cioffi, Lightweight geopolymer-based hybrid materials, *Composites Part B: Engineering* 128 (2017) 225–237. doi:10.1016/j.compositesb.2017.07.020.

31. R. Kumar, M. I. Ul Haq, A. Raina, A. Anand, Industrial applications of natural fibre-reinforced polymer composites – challenges and opportunities, *International Journal of Sustainable Engineering* 12 (2019) 212–220. doi:10.1080/19397038.2018.1538267.

32. C.-K. Park, C.-D. (Steve) Kan, W. T. Hollowell, S. I. Hill, Investigation of Opportunities for Lightweight Vehicles Using Advanced Plastics and Composites, *National Highway Traffic Safety Administration*. (2012) 416.

33. M.-Y. Lyu, T. G. Choi, Research trends in polymer materials for use in lightweight vehicles, *International Journal of Precision Engineering and Manufacturing* 16 (2015) 213–220. doi:10.1007/s12541-015-0029-x.

34. J. Hopewell, R. Dvorak, E. Kosior, Plastics recycling: Challenges and opportunities, *Philosophical Transactions of the Royal Society B: Biological Sciences* 364 (2009) 2115–2126. doi:10.1098/rstb.2008.0311.

35. O. Das, A. K. Sarmah, D. Bhattacharyya, Biocomposites from waste derived biochars: Mechanical, thermal, chemical, and morphological properties, *Waste Management* 49 (2016) 560–570. doi:10.1016/j.wasman.2015.12.007.

36. O. Das, N. K. Kim, A. L. Kalamkarov, A. K. Sarmah, D. Bhattacharyya, Biochar to the rescue: Balancing the fire performance and mechanical properties of polypropylene composites, *Polymer Degradation and Stability* 144 (2017) 485–496. doi:10.1016/j.polymdegradstab.2017.09.006.

37. O. Das, D. Bhattacharyya, D. Hui, K. T. Lau, Mechanical and flammability characterisations of biochar/polypropylene biocomposites, *Composites Part B: Engineering* 106 (2016) 120–128. doi:10.1016/j.compositesb.2016.09.020.

38. R. A. Mensah, V. Shanmugam, S. Narayanan, S. Mohammad, J. Razavi, A. Ulfberg, T. Blanksvärd, F. Sayahi, P. Simonsson, B. Reinke, M. Försth, G. Sas, D. Sas, O. Das, Biochar-added cementitious materials — a review on mechanical, *Thermal, and Environmental Properties* (2021) 1–27.

39. X. Jiat, B. Yan, Z. Hiew, K. Chiew, L. Yee, S. Gan, S. Thangalazhy-gopakumar, S. Rigby, Review on graphene and its derivatives: Synthesis methods and potential industrial implementation, *Journal of the Taiwan Institute of Chemical Engineers* (2018). doi:10.1016/j.jtice.2018.10.028.

40. P. W. Serruys, M. J. B. Kutryk, A. T. L. Ong, Coronary-Artery Stents, 354 (2009) 483–495. doi:10.1056/NEJMRA051091.

41. E. Salernitano, M. D. C. Migliaresi, Composite materials for biomedical applications: A review, *Journal of Applied Biomaterials and Biomechanics* 1 (2003) 3–18.

42. J. Rawles, S. Fialkova, Z. Xu, J. Sankar, Effect of alloying elements concentration and processing parameters on the structural and mechanical properties of lightweight magnesium alloys, *ASME, International Mechanical Engineering Congress and Exposition Proceedings (IMECE)* 3 (2021). doi:10.1115/IMECE2020-24598.

43. S. Verma, N. Sharma, S. Kango, S. Sharma, Developments of PEEK (Polyetheretherketone) as a biomedical material: A focused review, *European Polymer Journal* 147 (2021) 110295. doi:10.1016/J.EURPOLYMJ.2021.110295.

44. D. J. Kelsey, G. S. Springer, S. B. Goodman, Composite implant for bone replacement, (2016) 1593–1632. doi:10.1177/002199839703101603.

45. P. R. Monich, B. Henriques, A. P. Novaes de Oliveira, J. C. M. Souza, M. C. Fredel, Mechanical and biological behavior of biomedical PEEK matrix composites: A focused review, *Materials Letters* 185 (2016) 593–597. doi:10.1016/J.MATLET.2016.09.005.

46. A. Thien, N. K. King, B. T. Ang, E. Wang, I. Ng, Comparison of polyetheretherketone and titanium cranioplasty after decompressive craniectomy, *World Neurosurgery* 83 (2015) 176–180. doi:10.1016/J.WNEU.2014.06.003.

47. Chuan Silvia Li, Christopher Vannabouathong, Sheila Sprague, Mohit Bhandari, The Use of Carbon-Fiber-Reinforced (CFR) PEEK material in orthopedic implants: A systematic review, *Clinical Medicine Insights: Arthritis and Musculoskeletal Disorders* 8, CMAMD.S20354|10.4137/cmamd.s20354, (n.d.).
48. Ayesha Khalid, Romana Khan, Mazhar Ul-Islam, Taous Khan, Fazli Wahid, Bacterial cellulose-zinc oxide nanocomposites as a novel dressing system for burn wounds, *Carbohydrate Polymers* 164 (2017) 214–221. doi:10.1016/J.CARBPOL.2017.01.061.

Chapter 2

Metals and alloys for lightweight automotive structures

K. Ponappa and Yash Panchal

PDPM Indian Institute of Information Technology,
Design and Manufacturing, Jabalpur, India

CONTENTS

2.1 NEEDS OF METALS AND ALLOYS IN AUTOMOTIVE AREA

In 1885, the first two-seater car with a single-cylinder four-stroke gasoline engine, steel frame, three wire-spoked wheels, and the differential was developed by Carl Benz, as shown in Figure 2.1a.

From day one, steel was the primary choice for the structure and body of automobiles. During the initial time, steel competed with wood and aluminum as a body structure material. However, by the 1920s, steel was the material of choice because of its low cost, anticipated material property, and ability to form into complex shapes.

In the early 1900s, the idea of the body-on-frame design came into the picture by Ford, and in 1908 Ford introduced the Model T, as shown in Figure 2.1b. This vehicle had a steel load-bearing chassis, which supported all the mechanical parts.

DOI: 10.1201/9781003252108-2

(a)   (b)

Figure 2.1 (a) First two-seater car with a single-cylinder four-stroke gasoline engine [1]
(b) Ford Model T – 1908 [2]. (a) https://www.daimler.com/company/tradition/
company-history/1885-1886.html#:~:text=On1 January 29%2C 1886%2C Carl,1,
(b) https://corporate.ford.com/articles/history/the-model-t.html.

For the structure of automobiles and components under extreme heat conditions, the polymers are not the most viable solution, as they cannot withstand such high mechanical and thermal conditions, as most of them have a lower strength to cost than metal alloys with not so vastly available choice as metals. In such cases, metallic alloys and composites are the only feasible solution. With increasing demands and competition in the market, many new metals and alloys were introduced in the automotive industry to achieve identical quality within the equivalent price range.

2.2 DEMAND FOR WEIGHT REDUCTION

In 1967 the US government proposed corporate average fuel economy (CAFE) goals for the year 2000, with the primary objective being a reduction of fuel consumption as well as greenhouse gas reduction. One obvious and superior way is downsizing, but the downsizing of vehicle strategy has reached its limit. Because of this, many manufacturers have focused on the subsequent solution: the use of lightweight materials.

In the auto industry, the two major driving forces for innovation are cost and fuel economy, and light metals are the standard solutions in both cases. According to Global EV Outlook 2020, electric cars share only 2.6% of global car sales [3]. The weight of the electric automobile is much higher than gasoline automobiles, and the driving range of the electric automobile also depends on the overall weight of the vehicle. Driving range can't be improved by advancements in battery technology alone, but they must also be in conjunction with utilization of light metals and alloys.

Still, more than 95% of the market are traditional gas vehicles. The light-weight vehicle is the most efficient way to reduce greenhouse gas emissions and fuel consumption by powertrain downsizing, so all automobile manufacturers are seriously investing time and money in lightweight materials development. The impact of weight reduction on car performance is as follows: A 5%

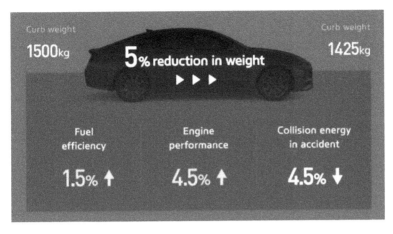

Figure 2.2 Effect weight reduction on car performance [4]. https://news.hyundaimotorgroup. com/Article/Hyundai-Kia-Automotive-Lightweight-Technology-Development.

reduction in vehicle weight will lead to a 1.5% enhancement in the fuel economy and a 4.5% increase in engine performance, as shown in Figure 2.2.

Weight reduction is ultimately related to carbon dioxide emissions, which means one kilogram of weight reduction reduces the carbon dioxide emissions by 20 kg [5]. Fuel economy depends on other factors, but vehicle weight is the most influential factor in city driving conditions. Weight reduction also has other advantages, e.g., a lighter vehicle has smaller brakes and engines. Along with weight reduction, the automobile should have enough weight to maintain the dynamics—the aim is to maintain strength with weight reduction. The natural inclination is to believe that "the heavier material will be stronger," but with advancements in material science, we now have more powerful and lightweight materials.

2.3 MATERIAL SELECTION

Traditionally, the "Lightweight" material term was assigned to Aluminum and Magnesium, as both materials are often used in the weight-reduction application of components. However, after more research and advancements in material science, two more materials, Titanium and Beryllium, have been added to this category. Weight reduction cannot be achieved by lowering the weight of a single component; for that, a comprehensive systems-engineering approach is required. As a result, it's no surprise that clean-sheet designs have achieved the most considerable weight savings. Table 2.1 shows the amount of weight savings that can be achieved by using lightweight materials, with the percentage increases in cost related to the material replacements compared with the low-carbon steel.

Table 2.1 Amount of weight saving and increase in cost using lightweight
materials when compared with low-carbon steel [6]

Material	Weight reduction (%)	Increase in cost (%)
Aluminum	40–50	130–200
Magnesium	55–60	150–250
Advanced high strength steel (AHSS)	15–25	100–150

2.3.1 Aluminum

Aluminum density (2.74 g/cm³) is 1/3 of the steel density (7.84 g/cm³).
The strength of aluminum will be improved using alloying elements and
heat treatments without much change in modulus of elasticity. Aluminum
alloys are preferred in many applications because of their light weight, easy
machinability, ability to produce near net shape products, and excellent
thermal conductivity [7]. Figure 2.3 compares the mechanical property of
aluminum alloys with the different steels.

With so many possible applications, in 2006 aluminum overtook cast iron
and took second place after steel in the list of the most used materials in
vehicles. With the accurate design, one kilogram of aluminum can substitute
the two kilograms of commonly used iron in the automotive industry. In the
early 1970s, 38 kilograms of aluminum were used in vehicles. This value
was increased to 212 kilograms in 2020, as shown in Figure 2.4.

Cast alloys, extruded alloys, and rolled alloys are the three types of aluminum alloys used in the auto industry. The chemical composition and properties of automotive alloys are used to classify them into series like 3000,
5000, 6000, and 7000, as shown in Figure 2.5.

Work hardenable 5000 series aluminum alloys such as 5182, 5454, and
5754 are supplied in the annealed (O) temper condition. Age-hardenable,
6000 series alloys such as 6009, 6022, and 6111 are provided in the solution
annealed and stabilized T4 temper condition. These alloys are used in aluminum alloy sheet manufacturing for automotive and light truck applications.
Table 2.2 lists the mechanical characteristics of aluminum body sheet alloys.
Each series contains many variations and subgroups based on the composition, production techniques, and tempers (ex. T4, T6) of various alloys.

The supply balance between 5000 series and 6000 series automobile sheet
alloys is shifting more and more towards 6000 series alloys, according to
Innoval Technology. The 6000 series alloys account for at least 80% of the
current amount supplied to automakers. While the 5000 series alloys have a
high strength-to-weight ratio, formability qualities, and full recycling compatibility, the 6000 series alloys have the dominant position since they are
adaptable, heat-treatable, highly formable, and weldable.

Each alloy series is designed for specific parts and places in automobiles,
as shown in Figure 2.6. Because of their resistance to strain marks in

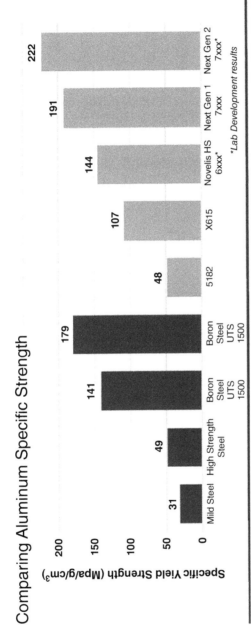

Figure 2.3 Comparison of strength between steel and aluminum alloys [8]. https://aluminiuminsider.com/aluminium-alloys-automotive-industry-handy-guide/.

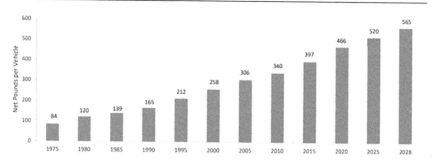

Figure 2.4 Growth of aluminum usage in vehicle industry over last 45 years [9].http://www.drivealuminum.org/wp-content/uploads/2017/10/Ducker-Public_FINAL.pdf.

New developments: high-strength aluminum sheet for body structures

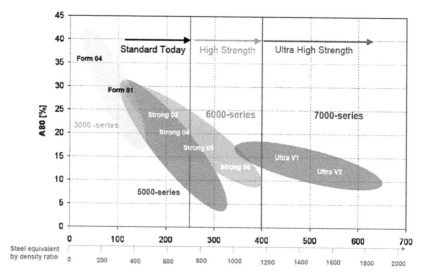

Figure 2.5 Comparison of different aluminum alloys [8].https://aluminiuminsider.com/aluminium-alloys-automotive-industry-handy-guide/.

stamping and greater strength, the 6000 series alloy sheet is widely used in exterior body panels, as seen in Figure 2.6. Internal body panels and body structure parts are more commonly made using the lower-cost 5000 series alloy sheets. Various monographs produced by The Aluminum Association include helpful information on design aspects, forming, connecting, and finishing of the aluminum alloy automobile sheet, as well as crash energy management and repair of the aluminum automotive sheet.

The introduction of aluminum matrix composites (AMCs), a fascinating category of innovative materials, was prompted by its following properties:

Table 2.2 The mechanical characteristics of aluminum body sheet alloys [10]

Alloy and temper	Ultimate tensile strength (MPa)	Modulus of elasticity (GPa)	Elongation (%) (in 50 mm length)	Yield strength (MPa)
5182–O	275	71	24	130
5454–O	250	70	22	115
5754–O	220	71	25	95
6009–T4	220	69	25	125
6009–T62	295	69	11	260
6022–T4	255	69	26	150
6022–T62	325	69	12	290
6111–T4	290	69	26	150
6111–T62	360	69	11	315

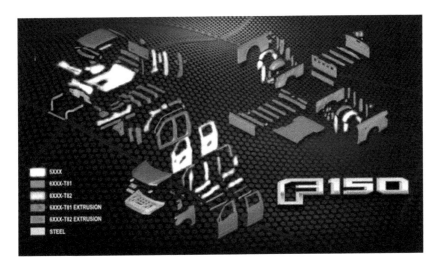

Figure 2.6 Different alloys for different parts of Ford F-150 pickup truck [11]. https://www.caranddriver.com/news/a15358912/in-depth-with-the-2015-ford-f-150s-aluminum-presented-in-an-alloy-of-facts-and-perspective/.

good ductility, superior strength-to-weight ratio, high strength and high modulus, excellent wear resistance, excellent corrosion resistance, high-temperature creep, low thermal expansion coefficient, and better fatigue strength [12]. They're frequently utilized in high-performance applications like automotive, military, industrial, electrical, and aerospace [12]. Ceramic materials such as SiC, Al_2O_3, B_4C, and MgO are commonly used to strengthen aluminum alloy matrices. Refractoriness, high compressive strength, high hardness, wear resistance, and other characteristics of these materials make them appropriate for application as reinforcement in composite matrixes [13].

A soft metal, such as aluminum, glides across strong metal, such as steel, without any external fluid or solid lubrication—it is expected that the aluminum will flow and cling to the steel, resulting in a low-shear strength contact. This theory is supported by the transfer of aluminum onto a steel ball during a standard ball-on-disk friction test [14]. Sliding will continue to transfer aluminum, and wear debris may develop due to the asperities of the hard steel ploughing the soft aluminum surface or flaking off portions from the transfer film [14]. Aluminum and steel couples have a high friction coefficient (0.5 to 0.6) [15]. The development of AMCs dispersed with solid lubricants is primarily aimed at overcoming aluminum's primary tribological disadvantages. Compared to the other matrix alloy, the friction coefficient of Al alloy-graphite composites is significantly lower.

2.3.2 Magnesium

With a density of 1.74 g/cm³, magnesium weighs 50% less than aluminum (2.74 g/cm³) and around 450% less than steel (7.84 g/cm³), becoming the lightest of all the family of engineering metals [16]. Table 2.3 lists the physical characteristics of magnesium, iron, and aluminum.

Many components in automotive applications must be ductile, and high energy absorption in accidents is a critical consideration. Optimizing the material's energy absorption is one direction in alloy and process development for wrought alloys. Other components, on the other hand, require more strength than ductility. As a result, the development of alloys is guided by a variety of criteria, and certain alloy groups can be found to give specific characteristics (Figure 2.7). Magnesium can be alloyed with manganese, aluminum, rare earth metals, zinc, thorium, or zirconium to improve the strength-to-weight ratio, making it useful in applications where weight reduction is needed. Due to this characteristic, along with cast iron and steel, copper-base and aluminum alloys have also been replaced by magnesium and its alloys [16, 17].

Magnesium also has castability, high ductility, and superior noise- and vibration-amplifying properties than aluminum [17]. Table 2.4 shows the advantages of using magnesium for any application.

The need for weight reduction of automobiles to obey the emission-control laws has reignited attention on magnesium [19]. Magnesium alloys

Table 2.3 The physical characteristics of magnesium, iron, and aluminum [16]

Characteristics	Magnesium	Iron	Aluminum
Density (g/cm³)	1.74	7.84	2.74
Elastic modulus (GPa)	44.126	206.842	68.947
Tensile strength (MPa)	240	350	320
Crystal structure	HCP	BCC	FCC

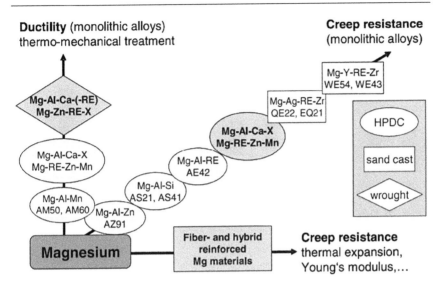

Figure 2.7 Magnesium alloy development guide [18].

Table 2.4 The advantages of magnesium [22, 23]

Property	Advantage
Machining	Its machining tools endure longer than aluminum, lowering expenses. The only drawback is that machining chips demand extra caution.
Specific stiffness	It has a higher specific stiffness than many polymeric materials and composites, allowing significant weight savings.
Specific strength	It has a specific strength comparable to cast iron and comparable to or greater than many typical automotive aluminum alloys, allowing more significant mass reduction than aluminum.
Fluidity	Its high fluidity enables exceptionally thin-walled castings (1.5 mm), which increases bulk reduction possibilities.
Damping	Its alloys offer superior damping properties as compared to other materials, which makes them appealing.
Hot formability	Using increased temperature forming methods, wrought magnesium may be produced into highly intricate forms.
Low-temperature properties	It lacks a brittle-to-ductile transition, allowing it to be utilized at extremely low temperatures.

have had a sporadic connection with the automobile industry, and since the 1920s, they've been utilized to reduce the weight of racing vehicles, allowing for faster acceleration and higher speeds—all for the sake of winning races.

Volkswagen was the first automobile industry to employ magnesium, which was in the 1950 Beetle model; each car had around 22 kg of magnesium [20]. Porsche experimented with magnesium engines for the first time in 1928 [21]. From 2005 to 2015, the average magnesium consumption and

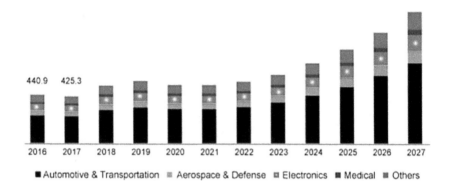

440.9   425.3

2016 2017 2018 2019 2020 2021 2022 2023 2024 2025 2026 2027

■ Automotive & Transportation   ▩ Aerospace & Defense   ▣ Electronics   ■ Medical   ■ Others

Figure 2.8 Magnesium and its alloys market growth in recent years [24]. https://www.grandviewresearch.com/industry-analysis/magnesium-alloys-market.

anticipated usage increase per automobile was 3 kg to 50 kg, respectively [18, 20]. For various vehicle parts in the past, aluminum and some plastics were the favored materials. Magnesium uses in the automotive industry have increased in recent years. Figure 2.8 depicts the recent increase of magnesium in automotive and other uses, as well as expected growth in the next five years.

2.3.3 Advanced high strength steel (AHSS)

Steel characteristics have advanced dramatically over the last century. Mild steel was introduced in the early 1900s; after that, high strength low alloys (HSLA) were introduced in the 1970s, and the first generations of Advanced High Strength Steel (AHSS) were introduced in the 1990s. [25]. The steel industry has developed new alloying and processing combinations in the last two decades to generate steel microstructures that provide better strength while reducing the size and weight of steel sections [26]. AHSS was developed to assist the automobile sector to achieve low weight requirements with its exceptional mix of high tensile strength and ductility, multiphase microstructures, and carefully chosen chemical compositions [27].

Although AHSS is not substantially lighter than conventional steels, its strength allows manufacturers to produce very thin gauges, decreasing vehicle weight. Strength-ductility performance, often known as "banana" diagrams, are widely used to depict the development of steel (Figure 2.9).

Steel is categorized into five types: mild steel, HSS, and first-, second-, and third-generation AHSS. The invention of dual-phase (DP) steels marked an evolutionary step forward, ushering in the age of microstructural engineering that was the start of the AHSS's first generation. The first generation of the AHSS family includes dual-phase (DP), ferritic-bainitic (FB), martensitic (MS), complex-phase (CP), and regular transformation-induced plasticity (TRIP). Multiphase microstructure of first-generation AHSS comprises

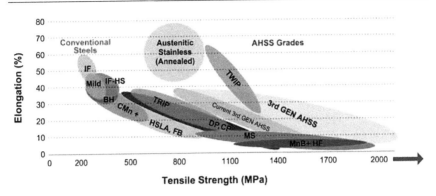

Figure 2.9 Steel classification for the automotive industry [25]. https://www.asminterna-tional.org/documents/10192/1849770/05370g_toc.pdf.

martensitic and ferritic phases for a balance of strength and formability, which accounts for better formability than the HSLA at the same strength level.

Specific heat treatments are used to generate the distinctive microstructure [25, 27]. Hot-formed (HF), twinning-induced plasticity (TWIP), and transformation-induced plasticity (TRIP) steels make up the second generation of AHSS [25]. First, two generations of AHSS are intended to fulfill the functional performance requirements of certain automobile parts [28]. The second generation has more formability than the first, but they are more expensive due to the high cost of alloying elements. As a result, the third generation of AHSS is now in the works. These steels are designed to have higher strength-to-ductility ratios and are projected to reduce structural weight by more than 35% [26].

As can be observed in Figure 2.10, mild steels are widely employed in car body structures and trunk closures, whereas HSS and AHSS steels have been used in sections of vehicles where energy absorption is crucial.

2.3.4 Titanium

Titanium and its alloys are low density, high strength, have excellent oxidation and corrosion resistance, and low modulus material. Titanium is considered one of the most plentiful metals on earth. As the fourth most abundant metal after aluminum, iron, and magnesium, titanium is still not widely used because of the difficulty in obtaining and mining it. Titanium has long been utilized in airplanes, saltwater desalination plants, heat exchangers, and electric power plants as a lightweight, robust, and corrosion-resistant material. Its attractive surface look and luxury feel have recently found growing uses in information technology and consumer products, such as athletic equipment. Titanium is suitable for a wide range of automotive applications and is already used in the aviation and aerospace industries.

Mild Steel
High-Strength Steel
Extra High-Strength Steel
Ultra High-Strength Steel

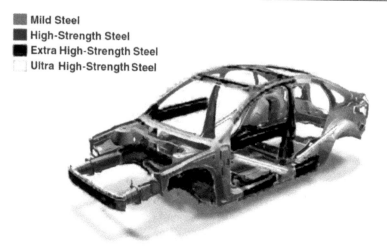

Figure 2.10 Distribution of steel in the automobile [29]. https://www.thefabricator.com/stampingjournal/article/stamping/lightweighting-still-dominates-great-designs-in-steel-seminar.

Commercially pure titanium is usually cold-formed. Tubes and sheets of titanium can be formed at room temperature similar to steel processing and with the same equipment. To achieve trouble-free production of titanium products, the following three points must be considered in detail: (1) Titanium's ductility is lower than that of iron and stainless steel. To achieve the critical low radius bending requirements, a larger bend radius may be required; (2) As titanium's modulus of elasticity is around half that of steel, titanium tends to spring back after forming. Because of this, over-forming should be done to compensate for this; (3) Titanium is prone to galling when used with non-lubricated forming tools. Tooling that is properly oiled and clean is usually recommended. The comparison of the material property of pure titanium and the most used titanium alloy Ti6Al4V is compared with other metals like iron, stainless steel, magnesium, and aluminum in Table 2.5.

Table 2.5 Comparison of property of titanium with other metals [30]

Material	Specific gravity	Young's Modulus (GPa)	Tensile strength (MPa)
Grade 2 pure Titanium	4.51	106	450
Titanium Alloy (Ti6Al4V)	4.43	114	900
Stainless Steel	7.95	200	600
Iron	7.84	207	350
Magnesium	1.74	45	240
Aluminum	2.74	69	320

The main focus of titanium alloy manufacturers is to create low-cost alloys, but it is apparent that low-cost titanium alone will not ensure automotive use; however, a competitively priced titanium component will as automotive manufacturers endeavor to reduce the cost of the entire system. Constructive growth has been made toward the cost-effective use of titanium in the sector of automobiles. Cold wound springs made of low-cost beta titanium alloy and exhaust systems made of commercially pure titanium are the most common uses. Both components are produced by titanium for the auto industry with tools and methods originally designed for steel parts fabrication.

Still, low-cost titanium component production processes must be established. Automotive designers and the industry also keep track of the cost-benefit of titanium and stay up to date on all design and production advances in titanium alloy in order to minimize manufacturing problems and costs, as well as to improve vehicle quality.

2.4 APPLICATIONS

Automotive experts have identified a variety of uses due to the unique characteristics of advanced lightweight materials. For each material, a few significant applications are listed below.

2.4.1 Aluminum applications

For the engine cylinders, cast iron cylinder liners are usually needed because of the poor wear properties of aluminum. Porsche, for the first time in automotive history, has used aluminum MMCs for cylinder liners by inserting a porous silicon preform into the cast aluminum block, while Honda performed a similar technique with die-cast aluminum bores by combining alumina and carbon fibers instead of porous silicon preform. Today, compared to cast iron liners, these aluminum MMCs enhance not only wear characteristics but also cooling efficiency. As discussed earlier, the lowest friction coefficient can be achieved using an aluminum-graphite composite. Functionally graded aluminum-graphite cylinder liners are produced using centrifugal casting, as shown in Figure 2.11.

Audi Space Frame (ASF), a revolutionary multi-material frame, was introduced in the 2017 Audi R8 Spyder V10. ASF weighs only 208 kilograms, and it is made possible via the combination of 79.6% aluminum and 20.4% carbon fiber-reinforced polymer components, as shown in Figure 2.12 (b) [31]. Further weight reduction was achieved by making the car's outer body parts like the door, roof, wing, hood, and liftgate using the aluminum sheet, as shown in Figure 2.12a.

Aluminum has been used in many luxury cars, but in 2015 Ford introduced aluminum in its best-selling pickup, the F-150; its aluminum body is

Figure 2.11 Functionally graded aluminum-graphite piston and cylinder liners [12]. https://www.asminternational.org/documents/10192/1902912/amp_17003p19.pdf/84deabbd-eae3-47a8-9c39-c1783d1414f8/AMP17003P19.

shown in Figure 2.13. One of the most concerning questions was how the truck's aluminum body and bed would fare in today's strength and safety criteria. According to the first car crash test result by the Insurance Institute for Highway Safety (IIHS), aluminum offers the same level of protection and strength as steel [33]. In recent years many car manufacturers have increased the aluminum content in vehicles—even the Tesla Model-S body is made up of an aluminum sheet. Figure 2.14 shows the amount of aluminum used in some cars in 2020 with the share in total production.

2.4.2 Magnesium applications

The quantity of magnesium utilized in the automobile sector is anticipated to rise by at least 300% over the next 8 to 10 years. Increasing the applications of magnesium alloy in each automobile will help the world meet its greenhouse gas reduction targets. Recent advancements in the manufacturing of magnesium alloys have expanded their potential for use in the automobile sector. The present batch of magnesium alloy automobile components is mostly produced using the above-mentioned casting techniques. To extend Mg's use in the automobile sector in the long run, more research into the forming processes of Mg alloys is required.

Most auto manufacturers use magnesium in the production of many parts of their vehicles. Parts like the engine block, steering wheel frame, seat frame, wheel rims, instrument panel, cylinder head, transmission case, clutch case, lower crankcase, intake manifold, cylinder block, air intake system, oil pump body, steering link bracing, camshaft drive chaincase, oil pump body, gear controls housing, and many more are being produced by car manufacturers such as BMW, Ford, GM, Toyota, Mercedes-Benz, Honda, McLaren Automotive, Volkswagen, Porsche AG, etc. Some of the applications are shown in Figure 2.15. A combination of aluminum and magnesium is also used for the engine block in which the outer casing is made using magnesium, while the inner part is made from aluminum.

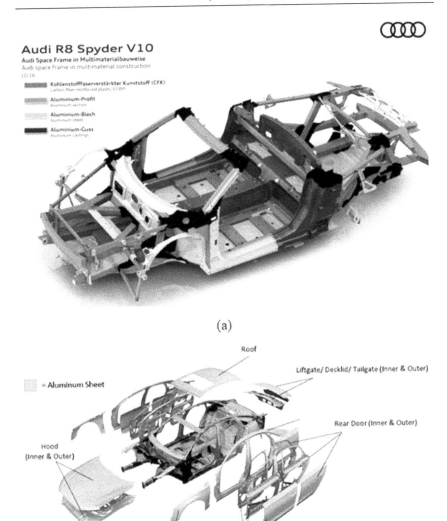

Audi R8 Spyder V10
Audi Space Frame in Multimaterialbauweise
Audi space frame in multimaterial construction
10/16

▨ Kohlenstofffaserverstärkter Kunststoff (CFK)
Carbon fiber-reinforced plastic (CFRP)

▨ Aluminium-Profil
Aluminium section

▨ Aluminium-Blech
Aluminium sheet

▨ Aluminium-Guss
Aluminium castings

(a)

Roof

Liftgate/ Decklid/ Tailgate (Inner & Outer)

▨ = Aluminum Sheet

Rear Door (Inner & Outer)

Hood
(Inner & Outer)

Fender / Wing

Front Door (Inner & Outer)

Product example by Audi

(b)

Figure 2.12 (a) Audi Space Frame (ASF) in the 2017 Audi R8 Spyder V10 [31] (b) Closure components made up of aluminum sheet [32]. (a) https://www.lightmetal-age.com/news/industry-news/automotive/new-audi-r8-spyder-features-multi-material-frame/, (b) https://1pp2jy1h0dtm6dg8i11qjfb1-wpengine.netdna-ssl.com/wp-content/uploads/2020/08/DuckerFrontier-Aluminum-Association-2020-Content-Study-Summary-Report-FINAL.pdf.

Figure 2.13 Ford F-150 (2015 Model) Aluminum Body [33]. https://www.greencarreports. com/news/1099466_2015-ford-f-150-aluminum-body-pickup-mixed-iihs- safety-scores.

2020 Vehicle Segment Average Aluminum Pounds and Share of 2020 Production:

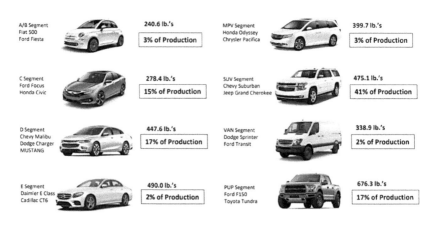

A/B Segment Fiat 500 Ford Fiesta	240.6 lb.'s	3% of Production
MPV Segment Honda Odyssey Chrysler Pacifica	399.7 lb.'s	3% of Production
C Segment Ford Focus Honda Civic	278.4 lb.'s	15% of Production
SUV Segment Chevy Suburban Jeep Grand Cherokee	475.1 lb.'s	41% of Production
D Segment Chevy Malibu Dodge Charger MUSTANG	447.6 lb.'s	17% of Production
VAN Segment Dodge Sprinter Ford Transit	338.9 lb.'s	2% of Production
E Segment Daimier E Class Cadillac CT6	490.0 lb.'s	2% of Production
PUP Segment Ford F150 Toyota Tundra	676.3 lb.'s	17% of Production

Figure 2.14 Amount of aluminum used in some of the cars in 2020 with the share in total production [9]. http://www.drivealuminum.org/wp-content/uploads/2017/10/ Ducker-Public_FINAL.pdf.

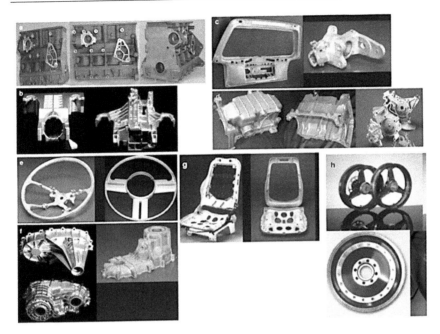

Figure 2.15 Magnesium components of car parts [34].

General Motors achieved about 50% weight reduction in the instrumental panel using magnesium, as shown in Figure 2.16.

2.4.3 AHSS applications

With the increasing use of light metals, composite metals, and polymers, the steel applications in the car are decreasing. At present, the majority of steel is found in the car structure only. Even after so many advancements in material science, steel is considered the best and most common material with which to manufacture a safe vehicle [36].

AHSS can absorb more amounts of energy during an impact. For that reason, AHSS is used in frontal and rear crumple zones in most cars. AHSS has replaced steel in passenger safety components such as A-pillars, hinge pillars, side-impact beams, B-pillars, waistline bumpers, reinforcements, roof heads, roof rail, door inner beam, and seats, as shown in Figure 2.17.

2.4.4 Titanium applications

Because of titanium's excellent heat and corrosion resistance, titanium has been replacing aluminum parts in the manufacture of commercial and military aircraft. Additionally, lightweight airplanes use less fuel. As a result, titanium is currently employed in a variety of aircraft components, including

(a)

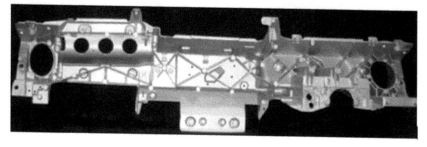

(b)

Figure 2.16 Instrument panel by GM (a) Cast iron instrument panel (12.8 Kg) (b) Current-generation Mg instrument panel (6.9 Kg) [35].

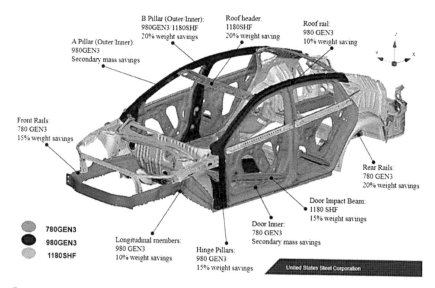

Figure 2.17 Car structure made up of steel [37]. https://www.thefabricator.com/stampingjournal/article/stamping/third-generation-advanced-high-strength-steel-emerges.

frames, landing gear, and fastening elements. Titanium is also used to create engine parts by aircraft manufacturers due to its ability to withstand extreme temperatures from sub-zero to 600° C. As a result, titanium is utilized in engine discs, shafts, casings, and blades, preparing engineers for its application in the vehicle industry. Some applications are mentioned below:

✓ Titanium is an excellent material for making connecting rods. Engine connecting rods made using titanium alloy can efficiently enhance engine quality, increase fuel efficiency, and reduce pollutants. When compared with steel connecting rods, titanium connecting rods can increase quality by 15% to 20%.

✓ The primary requirement of a valve spring seat is high fatigue resistance and strength. Heat treatment beta phase titanium alloy can achieve high strength by solution treatment aging, thus Ti-15Mo-3Al-2.7Nb-0.2Si and Ti-15V-3Cr-3Al-3Sn are more appropriate materials for valve spring production. Mitsubishi Motors uses the same material combination for valve spring seat production.

✓ Automotive engine valves made from titanium alloy have a longer service life, lower fuel consumption, and improve vehicle reliability. When compared to steel valves, the quality of titanium valves can be increased by 30% to 40%. The intake valve is mostly made of Ti-6Al-4V, while the exhaust valve is primarily made of Ti-6242S for the current vehicles.

✓ Low modulus of elasticity, corrosion resistance, and admirable fatigue properties of titanium alloys will be the most suitable solution to increase the service life of automotive springs. Ti-13V-11Cr-3Al and Ti-4.5Fe-6.8Mo-1.5Al alloys are usually used in spring manufacturing.

✓ A car's exhaust system makes extensive use of titanium rods. Titanium and its alloy exhaust systems not only enhance prolonged life, durability, and improve aesthetics, but they also improve quality and fuel burn efficiency.

✓ Turbochargers improve the combustion efficiency of engines, and they can also improve the power and torque of the engine as well. But for such kind of work, turbine rotors in turbochargers must operate for long periods in a high-temperature environment, thus the material used for such parts must be heat-resistant. But, because of their low melting point, light metals like aluminum cannot be utilized. While ceramic materials are generally used on turbine rotors because of their lightweight and good temperature resistance, their application is limited owing to their high cost and undefined form. This issue was first addressed by Tetsui, and as a solution, they created the TiAl turbine rotor. The result was unexpected, as it increases engine acceleration, along with an increase in durability and performance.

REFERENCES

1. Daimler, 1885–1886. The first automobile, (n.d.). https://www.daimler.com/company/tradition/company-history/1885-1886.html#:~:text=On January 29%2C 1886%2C Carl,1 (accessed September 19, 2021).
2. Ford, THE MODEL T, (n.d.). https://corporate.ford.com/articles/history/the-model-t.html (accessed September 19, 2021).
3. IEA (2020), *Global EV Outlook 2020*, IEA, Paris, 2020.
4. Hyundai Motor, Hyundai/Kia's Automotive Lightweight Technology Development, (2019). https://news.hyundaimotorgroup.com/Article/Hyundai-Kia-Automotive-Lightweight-Technology-Development (accessed September 20, 2021).
5. E. Ghassemieh, Materials in automotive application, state of the art and prospects, in: Marcello Chiaberge (Ed.), *New Trends and Developments in Automotive Industry*, Rijeka: IntechOpen, 2011: pp. 365–394. https://doi.org/10.5772/13286
6. M. Verbrugge, T. Lee, P. E. Krajewski, A. K. Sachdev, C. Bjelkengren, R. Roth, R. Kirchain, Mass decompounding and vehicle lightweighting, *Materials Science Forum*. 618–619 (2009) 411–418. https://doi.org/10.4028/www.scientific.net/MSF.618-619.411
7. N. Fatchurrohman, I. Iskandar, S. Suraya, K. Johan, Sustainable analysis in the product development of al-metal matrix composites automotive component, *Applied Mechanics and Materials*. 695 (2015) 32–35. https://doi.org/10.4028/www.scientific.net/AMM.695.32
8. D. Goran, *Aluminium Alloys in the Automotive Industry: a Handy Guide, Aluminium Insider*. (2019). https://aluminiuminsider.com/aluminium-alloys-automotive-industry-handy-guide/ (accessed September 23, 2021).
9. Ducker Worldwide, Aluminum content in North American light vehicles 2016 to 2028, 2017. https://www.drivealuminum.org/research-resources/ducker2017/ (accessed September 22, 2021).
10. J. R. Davis, P. Allen, S. Lampman, T. B. Zorc, S. D. Henry, J. L. Daquila, A. W. Ronke, Metals handbook: properties and selection: nonferrous alloys and special-purpose materials, *ASM International*, 2 1990.
11. S. Don, In-Depth with the 2015 Ford F-150's Aluminum, Presented In an Alloy of Facts and Perspective, Car and Driver. (2014). https://www.caranddriver.com/news/a15358912/in-depth-with-the-2015-ford-f-150s-aluminum-presented-in-an-alloy-of-facts-and-perspective/ (accessed September 23, 2021).
12. A. Macke, B. Schultz, P. K. Rohatgi, Metal matrix composites offer the automotive industry an opportunity to reduce vehicle weight, improve performance, *Advanced Materials and Processes*. 170 (2012) 19–23.
13. S. Debnath, L. Lancaster, M. H. Lung, Utilization of agro-industrial waste in metal matrix composites: towards sustainability, *World Academy of Science, Engineering and Technology*. (2013) 1136–1144.
14. A. D. Sarkar, J. Clarke, Friction and wear of aluminium-silicon alloys, *Wear*. 61 (1980) 157–167.
15. S. V. Prasad, K. R. Mecklenburg, Self-lubricating aluminum metal-matrix composites dispersed with tungsten disulfide and silicon carbide, *Lubrication Engineering*. 50 (1994).

16. G. Davies, 3 - Materials for consideration and use in automotive body structures, in: G. Davies (Ed.), *Materials for Automobile Bodies*, Butterworth-Heinemann, Oxford, 2003: pp. 61–98. https://doi.org/10.1016/B978-075065692-4/50020-0

17. C.-C. Jain, C.-H. Koo, Creep and corrosion properties of the extruded magnesium alloy containing rare earth, *Materials Transactions*. 48 (2007) 265–272.

18. C. Blawert, N. Hort, K. U. Kainer, Automotive applications of magnesium and its alloys, *Transactions of the Indian Institute of Metals*. 57 (2004) 397–408.

19. D. Eliezer, E. Aghion, F. H. (Sam) Froes, Magnesium science, technology and applications, *Advanced Performance Materials*. 5 (1998) 201–212. https://doi.org/10.1023/A:1008682415141

20. H. Friedrich, S. Schumann, Research for a "new age of magnesium" in the automotive industry, *Journal of Materials Processing Technology*. 117 (2001) 276–281. https://doi.org/10.1016/S0924-0136(01)00780-4

21. S. Schumann, The paths and strategies for increased magnesium applications in vehicles, *Materials Science Forum*. 488–489 (2005) 1–8. https://doi.org/10.4028/www.scientific.net/MSF.488-489.1

22. B. R. Powell, A. A. Luo, P. E. Krajewski, Magnesium alloys for lightweight powertrains and automotive bodies, in: *Advanced Materials in Automotive Engineering*, Elsevier, 2012: pp. 150–209. https://doi.org/10.1533/9780857095466.150

23. M. M. Avedesian, H. Baker, ASM speciality handbook: Magnesium and magnesium alloys, *ASM International*. (1999).

24. Magnesium Alloys Market Size, Share & Trends Analysis Report By Application (Automotive & Transportation, Aerospace & Defense, Electronics), By Region (MEA, North America, APAC), And Segment Forecasts, 2020–2027, 2020. https://www.grandviewresearch.com/industry-analysis/magnesium-alloys-market (accessed September 27, 2021).

25. M. Y. Demeri, Advanced high-strength steels: science, technology, and applications, *ASM International*, 1 (2013) 301.

26. A. I. Taub, A. A. Luo, Advanced lightweight materials and manufacturing processes for automotive applications, *Mrs Bulletin*. 40 (2015) 1045–1054.

27. Satyendra, Steels for Automotive Applications, Ispat Guru. (2015). https://www.ispatguru.com/steels-for-automotive-applications/ (accessed September 28, 2021).

28. Trefis Team, Trends in steel usage in the automotive industry, *Forbes* (2015). https://www.forbes.com/sites/greatspeculations/2015/05/20/trends-in-steel-usage-in-the-automotive-industry/?sh=5b508b061476 (accessed September 28, 2021).

29. B. Kate, Lightweighting still dominates Great Designs in Steel, FMA The Fabricator. (2018). https://www.thefabricator.com/stampingjournal/article/stamping/lightweighting-still-dominates-great-designs-in-steel-seminar (accessed September 28, 2021).

30. Y. Yoshito, T. Isamu, F. Hideki, Y. Tatsuo, Applications and features of titanium for automotive industry, *Nippon Steel Technical Report*. 85 (2002) 11–14.

31. New Audi R8 Spyder Features Multi-Material Frame, Light Metal Age. (2016). https://www.lightmetalage.com/news/industry-news/automotive/new-audi-r8-spyder-features-multi-material-frame/ (accessed October 5, 2021).

32. Ducker Frontier, 2020 North America light vehicle aluminum content and outlook, 2020.
33. E. Stephen, 2015 Ford F-150 Aluminum-Body Pickup: Mixed IIHS Safety Scores, 2015. https://www.greencarreports.com/news/1099466_2015-ford-f-150-aluminum-body-pickup-mixed-iihs-safety-scores (accessed October 5, 2021).
34. M. K. Kulekci, Magnesium and its alloys applications in automotive industry, *International Journal of Advanced Manufacturing Technology*. 39 (2008) 851–865. https://doi.org/10.1007/s00170-007-1279-2
35. A. Luo, Magnesium casting technology for structural applications, *Journal of Magnesium and Alloys*. 1 (2013) 2–22. https://doi.org/10.1016/j.jma.2013.02.002
36. Automotive Steel Processing: AHSS and Galvanized Steel, National Materials. (2020). https://www.nationalmaterial.com/automotive-steel-processing-ahss-and-galvanized-steel/ (accessed October 7, 2021).
37. D. Michael, Third-generation advanced high-strength steel emerges, *Stamping Journal*. (2017). https://www.thefabricator.com/stampingjournal/article/stamping/third-generation-advanced-high-strength-steel-emerges (accessed October 7, 2021).

Chapter 3

Polymers for structural applications

K. Arunprasath and M. Vijayakumar
PSN College of Engineering and Technology, Tirunelveli, India

Pon Janani Sugumaran
National University of Singapore, Singapore

P. Amuthakkannan
PSR College of Engineering, Sivakasi, India

V. Manikandan
PSN College of Engineering and Technology, Tirunelveli, India

V. Arumugaprabu
Kalasalingam Academy of Research and Education, Krishnankoil, India

CONTENTS

3.1 NATURAL FIBER AS REINFORCEMENT IN POLYMERS

Natural fibers make for good quality and strength in polymer composite materials. They do, however, lack in moisture-absorbing properties. Green showcasing, new mandates on reusing, social impact, and its changes have driven shoppers to change to materials that do not harm the ecosystem. The degree of individual and property assurances against the dangers in the combat zone and revolt circumstances has been created by the progression of the assaulting weapons.

Regular filaments can be categorized into two classes: creature-based strands and plant fiber. Casing silk, chicken quills, fleece, and insects silk are generally utilized as a fiber, fundamentally used for biomedical applications.

DOI: 10.1201/9781003252108-3

These natural fibers should be biodegradable, which implies the capacity to separate and to be ingested by the human body [1].

Suresh et al. [2] said that as of late, regular fiber-supported composites have been utilized in different applications because of their minimal expense, light weight, high solidarity-to-weight proportion, inexhaustibility, low thickness, and energy necessities for handling, and so on; in any case, with specific conditions and burden necessities, their constraints compel their application. To resolve the issue of regular filaments' substandard burden prerequisite and mechanical properties, fiber treatment and hybridization have gotten huge consideration by analysts. Using these strategies, specialists can plan a sandwich design of negligible water assimilation and ideal mechanical execution. Here, the possible utilization of regular fiber materials in a sandwich structure application is extensively examined. This audit will focus on the current investigations of sandwich structure materials and the potential to utilize normal fiber composites in sandwich structure applications.

The more extended the polymeric chain of the fiber, the better its similarity for use in materials specifically. Cellulosic strands have this property and have hydrogen bonds on the neighboring rings and phenyl rings on the spine. In addition, polymeric filaments have some limits, like a low firmness, which is responsible for giving low assurance against heat. At the point composites are produced using oil-based strands, ozone-harming substances are created, though cellulosic fiber-based composites could limit this test. Regular strands display awesome protection properties, which is why they can be utilized in the development and auto areas. There are not many advances accessible to create short, long, and consistent fiber fortifications, among which ceaseless strands are compounded with thermoplastic and another by thermosetting polymerization [3].

Both short and long fiber polymer composites have been applied in the automobile business. The construction of thermoplastics can be translucent, indistinct, or even semi-glasslike and is influenced by the diverse handling innovations of the polymers. Thermoplastic polymers are made by various strategies, for example, infusion trim, expulsion, and pressure shaping. Distinctive normal filaments are utilized for fortifications, alongside the grid, to upgrade the strength and execution of the composites.

Frequently, viable added substances are also added to upgrade the exhibition of composites. The properties of the natural fiber reinforcement polymer (NFRP) composites primarily rely upon the physical and compound nature of fiber support. The synthesis of the normal fiber (specifically, the non-cellulosic parts like hemicellulose, lignin, waxes, gelatin, and so forth) assumes the primary part in influencing mechanical strength [4].

Expulsion of these non-cellulosic segments upgrades the mechanical strength of the composite material, since their essence on fiber impedes the holding capacity of composite and its network. Normal filaments are for the most part hydrophilic, which decreases holding capacity. The NFRP composite's mechanical strength is then diminished because of this reaction. Yet, this can be overcome by the alteration of the fiber surface with synthetic treatment, which can be accomplished by the treatment of regular fiber with different pre-treatment techniques. The fiber surface can be adjusted by physical as well as substance treatment [5].

The composites made from glass fiber or carbon fiber support with very good strength for composite materials. For economic improvement, there is an extraordinary need just as a challenge for each industry to supplant the non-supportable products with maintenance. Attributable to the above reasons, it prompts the supplanting of engineered filaments with regular strands as support in fiber-built-up composites. Other than this, normal filaments likewise have numerous different advantages contrasted with manufactured filaments—for example, low relative thickness, minimal expense, high effect obstruction, and high adaptability, low explicit gravity, less abrasiveness to gear, less well-being perils, measure agreeable, lower nursery emanations, recyclability, and CO_2 non- reaction [6, 7].

The use of natural fiber-supported polymer composites (NFPC) in the transportation industry needs a stronger material because of lightweight, unique properties, less expense, and more useful items. In any case, the primary weaknesses of these filaments are their poor dimensional strength and high hydrophilic nature. Interfacial holding between the fiber and grid assumes an essential part in choosing the mechanical qualities of composites. Different synthetic fixes are applied for upgrading the fiber-grid grip, which results in better mechanical attributes of the composites. The aviation and car industries want to transition from traditional materials (which are high thickness) to composite materials to diminish the general load of the vehicle [8].

Because the heterogeneous idea of NFRPs varies from homogeneous materials like metals and polymers, some deformities occur when handling NFRPs through customary cutting strategies such as high surface harshness and material harm at the cutting zone. To eliminate these difficulties, new cutting strategies were considered. Offbeat cutting strategies didn't take into account the impact of cutting powers, which are the primary driver of cutting imperfections in conventional cutting cycles. The most unmistakable eccentric cutting cycles are grating waterjet and laser bar cutting advancements, which are used for cutting different NFRPs. Normal composites and other fiber-built-up polymer FRPs have numerous comparable properties as they are produced using similar polymers in the production of both NFRPs and other FRPs, so part of the cutting cycle of FRPs is discussed to help this examination. FRPs comprise aramid, glass, and carbon and are utilized to build up polymers [9].

Supplanting engineered FRP with normal FRP has other thorough benefits as far as the natural effects, degradability, and, furthermore, recyclability. The carbon impression of assembling one-ton flax fiber is approximately 35% lower than that of assembling one-ton glass fiber. Hemp fiber composites were found to emanate 10% to 50% less ozone-harming substance when contrasted with glass fiber partners. Fatigue cycle investigation reasoned that the creation of reed strands burns just 7.5% energy, 20.6% bio-compound oxygen, and discharges 32.3% carbon dioxide when contrasted with equivalent weight glass filaments. The applied life-cycle assessment (LCA) evaluated the natural effects of flax FRP, and glass FRP printed circuit sheets (PCB) were considered; the fundamentally lower ecological effect was accomplished by flax FRP [10]. Kumar et al. [11] said that among the polymer composites, those fused with filaments were considered the most significant attributable to their elevated strength as well as firmness on a weight premise.

An outstanding development also connected with this unique gathering of fiber-built-up polymer composites is that, during the 1960s, a few superior manufactured strands, for example, aramid, super high atomic weight polyethylene, and Poly-p-phenylene-2, 6-benzobisoxazole (PBO), were created and have been applied as support in polymer composites. Be that as it may, with the ever-increasing population of the planet—which necessitates practical solutions to forestall contamination and availability of non-sustainable assets—the substitution of manufactured materials with natural ones is an ongoing critical issue.

In polymer composites natural fibers are used as reinforcement materials, and the results have been positive. Owing to its, origin, it can behave based on the chosen applications. Major areas such as marine, mechanical, aerospace, and automobile engineering are correctly utilizing the functional approaches of NFRC for various lightweight applications. Figure 3.1 represents the various assessment of NFRC.

3.2 SYNTHETIC FIBER AS REINFORCEMENT IN POLYMERS

All over the world there is a call for institutions to "practice environmental awareness." Engineered fiber-built-up composite materials can cause ecological concerns, which are better handled by natural materials. Natural assets contribute considerably to the development of the output of any growing nation where the use of these materials creates jobs for people in provincial areas [12]. The mechanical properties of synthetic fibers are mainly affected by the enormous scope of boundaries like volume of strands, filament length, strands' side quantitative connection, fiber-network grip, fiber direction, which has higher stress at the interface.

Consequently, to work on the, generally speaking, mechanical aspects of composites, the properties of the network and strands must be looked at first. A few examinations are made on various regular filaments like jute and bamboo to check the aftereffect of those filaments on the mechanical and actual strength of composite materials. Better connection at the interface between the strands and thusly lattices has higher mechanical conduct in the

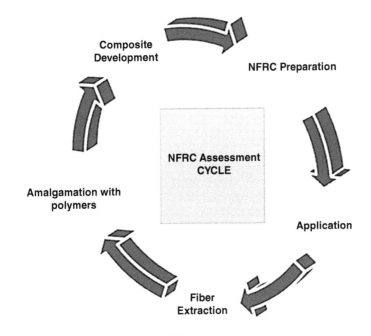

Figure 3.1 Various assessment cycles of NFRC.

composite. It has been confirmed that the mechanical properties of composites improve with an increase in interfacial strength [13].

Velram Balaji Mohan and Debes Bhattacharyya [14] discussed conductive glass fiber yarns with the utilization of Carbon Nano Tubes (CNT) covering electrophoretic deposition (EPD) in which a homogeneous covering of CNT was finished on the glass fiber, yet the manufacturing cycle included different strides before arriving at the last state EPD, and the conductivity was simply ~ 1×10^{-4} S.cm^{-1}. Adding glass and jute fiber changed the mechanical properties of the composites. But after adding,CNT, the composite combination showed better surface stability. Once more, the cycle vigorously included a manufactured substance affidavit that included wet synthetics while electrical conductivity was accomplished distinctly as 6.67×10^{-5} S.cm^{-1} with single glass fiber yarns; the jute fiber yarns even had a lot of lower electrical conductivity. Glass filaments were inundated in CNT scattering in the controlled affidavit of CNT particles to make a conductive layer brought about worked on mechanical properties, yet showed poor electrical conductivity simply up to 1×1^{-10}. By and large, for making a covering layer on the fiber surface, it is fundamental to reinforce the interphase where the various leveled structure develops unequivocally and electron shipping channels are adequately made.

These projections in synthetic fibers are considered an unpredictable surface morphology, which upgrades interlocking at the fiber-framework interface and subsequently works on mechanical properties of the composites. In this, it is worth noting that the surface state of a hydrophilic normal fiber plays a significant part in anything but a hydrophobic polymer framework. To work on this surface attachment, a few therapies utilizing soluble base, silane, acetylation, and radiation, among others, have been examined; specifically, they proposed a layer-by-layer nano-engineered procedure to alter the outside of regular strands and produce interphases fit for improving mechanical properties of polymer composites. Late works utilized graphene-based materials as an answer for the regular fiber lower attachment to a polymeric network. These new strategies for surface adjustment are currently being considered in our ongoing exploration for application in tucum fiber. Nonetheless, in the current work, no surface treatment was performed to prevent extra cost manufacturing cost [15, 16].

Asim mohammed et al. [17] discussed that delamination is a significant disappointment mode for all of the fiber-based polymer composites after the application of the load. After the post-impact condition of kenaf/glass, cross-breed composite shows different results for low-velocity impact test. The results of the impacts test showed that the pressure harm diminished as the effect energy expanded. It was additionally found that the kenaf/glass mixture composite with a 25% kenaf fiber weight proportion provided results similar to those of a glass overlay composite.

The heat temperature properties influence the morphology of synthetic fiber-built-up composites altogether. Contingent upon the temperature, the composites extend or contract, making breaks in the composites that at last hold moisture. The normal filaments absorb dampness through capillary activity, causing thickness growing in composites. Cellulose-supported polyethylene composites were utilized to go through the oxidation interaction, which worked on the interfacial holding of cellulose and grid and positively affected warm debasement. Another trial on untreated and silane-treated sisal strands was done, and debasement temperature and mass misfortune were determined. The mass deficiency of 5%, 10%, and a half was recorded at 83, 255, and 360°C for untreated fiber; also, at 100, 278, and 365°C for saline-treated fiber in the expressed request. The hemicellulose of sisal strands was thermally debased at 297°C, and cellulosic content was corrupted as uncovered in the second stage at 365°C [18].

Since the synthetic fiber areal masses of the channels included little varieties inside the three comparator gatherings, the strength is analyzed utilizing the worth of the channel extreme pressure strength (N) per unit fiber areal mass (gsm), the strength efficiency (N/gsm). While the channels had different relative amounts of epoxy material, this strength efficiency was chosen to straightforwardly look at different sorts of fiber fortifications. The strength efficiency esteems are organized, a straight pattern for every fiber type.

This information shows that for every fiber type, strength efficiency expanded with composite thickness. This is a notable result for slender-walled underlying areas in pressure, where more slender segments are less efficient since they are powerless to pressure clasping of the dainty plate components of the segment. This information moreover demonstrates that carbon fiber composites have better strength efficiencies than any remaining fiber types, despite being similarly dainty, while glass fiber composite strength efficiencies are tantamount to natural fiber composites [19].

Engineered strands are man-made because they go through different cycles before turning into a fiber that expels filaments' building materials through spinnerets into air and water, fostering a string. Airplane parts, cars, building boards, and other elite items are fabricated utilizing this material. The most mainstream engineered filaments that are by and large utilized broadly in industry incorporate glass, carbon, and aramid [20].

Ganesh R. Chavhan and Lalit N. Wankhade [20] said that fiber metal composites that are manufactured with glass fiber, hardened steel 316 L, and aluminum combination AA1050/epoxy tested for effect, twisting, and elasticity. It was seen that fiber metal composites with hardened steel on the external area showed an improvement in harm resilience limit, energy retention, and firmness—contrasted with fiber metal composite installed with aluminum alloy. The damping impact of plain composite, shape-memory combination installed composite, and steel-installed glass fiber supported a plastic composite pillar. The logarithmic root technique was utilized to gauge the damping proportion. From the test results, it was noted that the damping proportion for shape memory combination half breed composite pillar was higher as contrasted with other composite covers.

Synthetic fiber that resulted from impacting the properties of the composite kenaf strands provides added benefits such as low handling cost that justifies modifying the manufactured strands. It was determined that the properties of the composites can be adjusted by altering the kenaf strands utilizing substance treatment. This way the regular filaments could display either thermoset or possibly thermoplastic aspects to improve the physical properties of composite material. The incorporation of synthetic fiber in the composite content supports the system to achieve better results apart from other environmental considerations [21]. Figure 3.2 shows the benefits of using synthetic fibers as a reinforcing material in the composite.

3.3 PARTICULATE AND FILLER AS REINFORCEMENT IN POLYMERS

Natural fiber-built-up composites have become more mainstream among buyers due to the richly accessible, eco-accommodating nature and minimal expense of the crude materials for some modern applications. The crack and effect properties of Kevlar built-up epoxy polymer composite are prized by numerous specialists. Other studies determined that hybridized composites containing manufactured and normal filaments show unrivaled

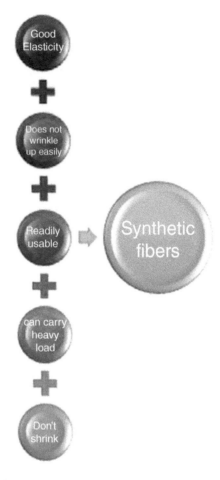

Figure 3.2 Benefits of using synthetic fibers as a reinforcing material in the composite.

tribological and mechanical properties when one of the supports is a polymer. Regardless, their investigation was restricted to only a certain gathering of filaments [22].

Sharma et al. [23] said that the predominant disappointment modes noticed because of drop molded fillers were mostly the same as that of round ones. Regardless, the extra disappointment mode for this situation was filler breakage, which adds to expanded energy scattering and relatively better crack durability. Among the three types of filler shapes, the bar molded fillers displayed an improved crack strength by filler pullout. This improvement in strength can be ascribed to the expanded energy dispersal by the pole-formed fillers due to a few distinguished filler breakages.

Varieties of shading contrast esteem in five equivalent segments of the influenced profundity for various blends after 200 h of UV openness with taking point 0 h as the outside of ex-acted tests and point 5 like the beginning of the unaffected zone. After UV openness, the shade of flawless sap became yellow and afterward earthy-colored appearance corruption of perfect epoxy gum. Also, it was seen that reduced essentially with expanding percent-times of fillers, which plainly shows that the fillers limited the shading contrast of the polymer tests presented to UV. For in-position, the shading changes of F40 and F60 tests are approximately 10% along with its influenced profundity, yet is over 30% for F20 at the uncovered surface [24].

Desire for thermoplastics is exploding, as the lightweight material for auto applications and other business purposes prompt more examination into the accessible polymer assets. In this examination, the idea of improving the quality of reused squander plastics as polymer-based composites was inspected. A particulate snail shell was acquired by establishing and sieving snail shells to get 53–63 µm passing, which was utilized as support in the reused squander plastics. Composites were created by adding changing extents of the snail shell particulate utilizing a haphazardly scattered interaction in a hot pressure forming machine kept at 190°C for 7 min [25].

Aggregation of filler particles during composite preparation makes an impediment that prompts their non-homogenous dissemination, preparing issues, without surface quality. The molecule associations are a vital highlight to be thought about on the advancement of another surface treatment or preparing strategies that can add to the improvement of homogeneity [26].

Synthetic fibers are likewise preferred in FRP composites because of their potential advantages like higher elasticity, light weight, warm recyclation, and duration and are being used in assembling different complex designs in composite enterprises that incorporate the aviation, marine, and auto industries. Filler materials can be characterized as latent material that can lessen the expense of the material, work on the material execution, and upgrade the mechanical properties of the composites. Normally, fillers are preferred alongside framework to get the great surface completion, which, in any case, would result in coarse construction and accordingly influence the mechanical properties of composites. Hard filler materials improve the tribological properties by lessening wear volume deterioration and wear rate. The regular filters can be utilized to build the general presentation of cross-breed composites [27].

Presto et al. [28] said that the nanoscale dynamic reaction of silica-filled elastic under strain is directed by the filler network structure and affects the naturally visible properties. Utilizing heterodyne X-beam photon relationship spectroscopy (HD-XPCS), in which the dispersing from the stressed elastic and a static silica-filled example is blended, extricated the total speeds of the silica particles during stress unwinding. As of late, X-beam photon relationship spectroscopy conducted on silica-filled ethylene propylene diene monomer elastic with silane coupling agent at 1 vol% filler stacking has noticed powerful anisotropy. In a particular area of the composite, the expansion of silane coupling specialists incredibly eases back molecule movement toward the applied strain.

Thermoplastic polymers are repetitively liquefied at a raised temperature and formed into the ideal shapes and are either semi-translucent or indistinct. Undefined thermoplastic polymers like ABS, acrylic, polystyrene, polyvinyl chloride, and so forth comprise arbitrarily situated atomic chains, though semi-translucent polymers like polyethylene, polypropylene, nylon, and so on display the changing level of crystallinity that controls the physical and mechanical properties. The translucent and nebulous polymers act diversely at the raised temperature. Shapeless polymers don't portray the genuine liquefying, yet relax to a state of a thick stream. The synthetic polymers have wide similarities with customary strands like glass and carbon, require low handling energy, are efficient, have low thickness, and offer simplicity of manufacture [29].

To build a sustainable society and help preserve the earth's natural resources, researchers must resolve the problem of increasing polymeric material waste. Of the several bio-based polymers in use, poly (lactic acid), commonly known as polylactide (PLA), is a biodegradable polymer widely used in automation, packaging, and 3D printing for lightweight applications. The ease of availability and composability makes PLA an excellent choice for a polymeric material; however, it lacks good thermal and mechanical stability. PLA degrades at approximately 400°C of dampness were seen. However, because of its low thermal stability, recycling the material and recovering useful components from the matrices is difficult. In addition, degradation is accelerated due to the presence of moisture and lactic acid and metal catalyst residue [30]. Figure 3.3 illustrates the benefits of using fillers/particulate as reinforcement in composite polymer composites.

3.4 ISSUES WITH POLYMERS IN COMPOSITE MATERIALS

Polyaniline and polypyrrole are acquired as powder for composite production. The in-situ surface manufacturing takes into consideration for the readiness of the composites with natural or inorganic segments with imminent application potential. [31].

Amongst biopolymers from creature sources, keratin is one the most prevalent, with a significant volume from side stream items from the steer, ovine, and poultry industries, offering numerous outlets that provide cost-effective and manageable developed materials. While numerous studies have examined the utilization of keratin in polymer-based biomaterials, little consideration has been paid to its potential in relation to other polymer frameworks. To that end, we present a survey that compares keratin's similarity with other manufactured, biosynthetic, and regular polymers and its effect on the materials helps to modify the setting of keratin used for construction, substance toolset, and techniques for extraction [32].

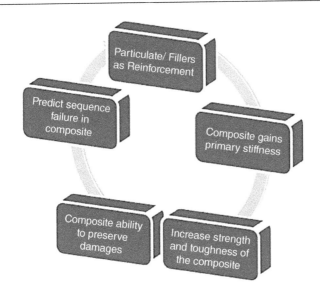

Figure 3.3 Benefits of using fillers/particulate as reinforcement in composites.

Tian et al. [33] said that most materials in use today are polymers, and later progress in innovative human advancement have generally included polymers. The technology and science regularly make progress in improvement of polymers, which has an extraordinary effect on advancing social progress. Likewise, the physical and compound properties of α-, β-and γ-CD are unique. The upper edge of the compact disc is in the scope of 0.45–0.77 nm; the lower edge is in the scope of 0.57–0.95 nm; the stature is 0.78 nm. With the progression of innovation, it has been discovered that the hydroxyl bunches could respond with various gatherings to deliver uncommon properties.

Natural fiber composites likewise have heat corruption issues, as the composite overlay is prone to overheating. Fiber treatment is one of the strategies that can be thought of to work on superficial level geology of filaments for mechanical employments. Accordingly, NFCs can satisfy a more extensive range of uses. There are three primary fiber medicines for working on the nature of NFCs: antacid, silane, and acetyl medicines. Antacid treatment, or mercerization, helps the evacuation of lignin and hemicellulose, though silane treatment gives a covering layer on the outside of the regular strands. For acetylation, mercerization eliminates the hemicellulose, normal fats, lignin, and waxes from the acetylating cellulose. In this instance, the interaction can yield a higher capability of hydroxyl gatherings and other responsive utilitarian gatherings on the filaments' surface [34].

The structure, properties, and quantity of layers of the graphene-based materials are identified with the method chosen for their creation. These cycles can be isolated into two principal classifications: the hierarchical and the granular perspectives. The hierarchical methodology combines several assembling techniques, which break down construction into nuclear layers, for example, mechanical peeling, graphite intercalation, cutting of carbon nanotubes (CNT), a decrease of graphite oxide, electrochemical peeling, among others. The granular perspective comprises techniques utilizing an elective wellspring of carbon as a structural block. The development of metal-carbon castings, epitaxial development in silicon carbide (SiC), and synthetic fume statement are examples of the base-up approach. Each strategy results in different outcomes that may be reasonable for specific applications. Many of them involve significant expense, which makes their application unreasonable [35].

The choices around planning a lightweight part for aviation applications are made dependent on different limitations such as cost, manufacturability, and so on; one hard breaking point is a strong necessity in underlying parts. This security factor represents a coincidental in-administration load that surpasses the configuration limit and doesn't represent varieties in mechanical properties, which are managed under various guidelines [36].

The complex polymer substance responses included in polymerization handling, it is surely known that emotional pressure bringing down and temperature motions as in space climate regard to standard conditions may genuinely influence the actual solidness, cause the event of outgassing marvels. Specifically, the mechanical test ought to give data about the compound idea of the removed gases, since these last-mentioned by re-buildup may prompt hurtful impacts in aviation missions where saving close by surfaces and touchy gadgets from defilement addresses an essential limitation [37].

The structural constituents, relative concentrations, and various fiber parameters like length, distribution, diameter, and orientation strongly influence the properties of the fiber composite. However, handling of an effective load transfer between the fiber and matrix requires that the interactions of the constituents should be strong. The chemical modifications were adopted for overcoming this issue and effect impactful influence on the properties of the composite. For banana fiber composites, the chemical treatments performed include alkaline, acetylation treatment, silane treatment, permanganate, benzoylation treatment, and maleated coupling reagents. However, all these chemical treatments have their limitations [38].

Ray et al. [39] said that pace of water assimilation relies upon the crude materials' state, nature of the fiber constituents, and ecological conditions such as temperature and dampness. The impact of manufactured fiber is just as normal fiber loadings to the composite on the level of water assimilation of various composites and hybrid composites. At last, the water retention conduct of various composites counting virgin PP is explored. It is seen that the level of water retention step by step increments with the joining of various filaments with various creations.

The properties of composites at cryogenic temperatures are not the same as those at typical temperatures, including the mechanical tests and other

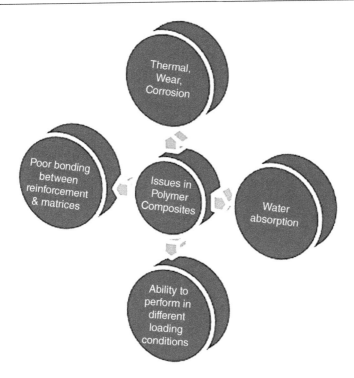

Figure 3.4 General issues in polymer composites.

practices. The polymer composite would become denser and weak with the lowering of surrounding temperature; subsequently, the general solidness and strength of CFRP materials will be fundamentally adjusted. On account of low temperature, the framework becomes fragile because of the lessening of sturdiness. It will move shear power to strands with shrinkage [40]. Figure 3.4 represents some of the issues in polymer composites.

3.5 APPLICATIONS

Fortified with exceptionally low rates of nanometric particles of around 2%–5%, polymer—when contrasted with the base tar—offer gigantic upgrades in terms of thermomechanical properties, boundary properties, and imperviousness to fire. Also, they can beat customary fillers and strands; for example, CB, glass, mineral fillers, calcium carbonate, graphite, metal oxides as far as warmth obstruction, dimensional dependability, and electrical conductivity. High CB focuses can deliver high thickness during preparing and debilitate mechanical performance. Furthermore, the toughness of the last article makes CPC an alluring option for applications customarily saved for metals [41].

Alsubari et al. [34], however, said that fibers have disorders for some load necessities, and this weakness has been countered through fiber treatment and hybridization. Sandwich structure, again, is a blend of two or more individual segments, which, when combined, provides better strength. Sandwich structures are utilized in a wide variety of modern material applications. They are known to be lightweight and acceptable at absorbing energy, giving predominant strength and solidness to weight proportions, and through plan joining, eliminate some segments from the center component. Today, numerous businesses utilize composite sandwich structures in manufacturing. Be that as it may, the use of normal fiber composites in sandwich structures at present is minimal.

Ceramics like alumina and zirconia are normally used for bone substitutions, inserts, platforms for tissue designing, Abandoned strength in physiological media, absence of controlled what's more, supported medication discharge, harm of sound tissues or cells due to powerless objective particularity, and so on, are a portion of the limits of MXenes utilized [42].

Regular fiber composites have enjoyed a renaissance over the last twenty years because of the need for eco-accommodating, biodegradable, and recyclable materials. They are currently being broadly utilized in ordinary items as well as in auto, bundling, sports, and development ventures. Hemp fiber is utilized in most of these items because of its unrivaled mechanical properties. Like other natural strands, hemp filaments require adjustments to improve their properties and bonding with polymer grids, and to decrease their hydrophilic aspects.

These adjustment techniques can be categorized into three significant classes: compound, physical, and natural. Synthetic techniques utilize substance reagents to minimize the hydrophilic tendencies of natural strands and to subsequently fuse with the grid. This chemical process strengthens the overall material without widely changing the compound structure of the filaments. They are cleaner and more straightforward than the substance approaches. Organic techniques utilize natural properties like growths, chemicals, and microscopic organisms to alter the fiber surface aspects. These strategies are not poisonous like substance techniques or energy-concentrated like physical strategies [43].

Electrochromic particle gels like ion gels, or polymer-based ILs with the possibility to modify their optical properties absorbance and conveyance upon the use of an outside electrical improvement for gadget advancement have moreover acquired exceptional attention [44].

Chemotherapy and radiotherapy is used to treat cancers, but an unfortunate side effect of such treatments is that healthy cells are negatively affected as well as the diseased ones. Tumor cells have low pH contrasted with healthy cells; consequently, the improvement of upgrades responsive materials delivered a solution for the downsides of the current anticancer treatment. These materials are initiated within the sight of catalysts, temperature, or pH. The presence of adversely charged hydroxyl, or on the other hand, fluorine bunches on the MXene surface, delivers simple electrostatic cooperation with the medication atoms that are emphatically charged [45].

The crown release is an air plasma type produced by an electrical release at the utilization of high voltage to sharp-tipped anodes, at low temperature and environmental pressing factor, which regularly includes oxygen-containing species. This is utilized in normal filaments adjustment, as it changes the surface energy of filaments, advances the arrangement of receptive locales onto the surface of the strand. Studies performed on hemp filaments adjusted by Corona releases that were hence utilized in PP-based composites have shown Young's modulus worth of the composite example expanded by 30%. Great outcomes were likewise acquired when normal tars were utilized as networks for hemp strands treated by corona discharges [46].

The incorporation of functional nanoparticles into a polymer matrix can effectively combine the respective advantages of each component, thus creating polymer composites with not only outstanding processability but also versatile functionalities. Filling the electrically conductive nanoparticles, such as carbonaceous or metallic particles, into the insulating polymer matrix, the polymer material can achieve an insulator conductor transition when a percolated conductive pathway inside the polymer matrix is formed to generate the conductive polymer composites.

According to their electrical resistivity, the CPCs can be used as antistatic or anticorrosive materials, electromagnetic interference (EMI) shielding materials, and conductors. Interestingly, it was found that the conductive network in a specific polymer matrix had a response to the external environments, such as mechanical stress, temperature, solvent, vapor, and so on. Since the charge transportation highly depends on the conductive network, tiny changes in the external environments may result in the change of the electrically conductive network, thus causing a remarkable change in the electrical resistance of the CPCs. Therefore, this stimuli-responsive behavior of the CPCs can be utilized to design highly sensitive sensors to detect or monitor the change in the external environments [47].

Conformity with lightweight RF materials is basic to incorporated radio wires for cutting-edge unmanned aerial vehicles (UAVs), as well as automated ground vehicles. The goal is to utilize smaller receiving wires that hold their frequency execution despite their smaller electrical capacity.

The airframe and stage require new non-conventional materials that are likewise low risk. New material properties like similarity, lightweight and solid shear, and malleable pressure appraisals are basic to the underlying incorporation also. Polymers and polymer-clay composites are shown to be truly adaptable and low risk; subsequently, appropriate for load-bearing capacity with the permittivity going from three to more than 13 with low risk digression less than 0.02 at a few GHz frequencies. Such composites can likewise be blended with attractive powders to accomplish higher penetrability, which is basic to radio wire scaling applications [48].

There is a growing need for materials harmless to the ecosystem that can lower the cost of conventional filaments. Cellulose is a whiz polymer material that is utilized as filler for the support of polymer materials. Microcrystalline cellulose is a part of cellulose. The objective of this examination is to create and describe polypropylene and microcrystalline cellulose-based composite. A silane surface adjustment procedure was utilized to change the cellulose surface, which was blended with polypropylene materials with various compositions. Cellulose is generally found in plants, wood, some marine creatures (e.g., tunicates), as well as parasites, microbes, invertebrates, and one-celled organisms.

Cellulose is the most abundant biomaterial on earth, which is inexhaustible and biodegradable. The sub-atomic makeup of cellulose is of prime significance as it clarifies the particular properties of cellulose, such as hydrophilicity, biodegradability, and high usefulness. Cellulose primary progression can be utilized to get ready high-strength small particles. Besides, the substitution of customary composite material by small composite materials has developed quickly during the past timeframe to conquer the impediments of conventional composites [49].

The examination is performed to comprehend the non-performance conduct of Continuous Fiber Reinforced Additive Manufactured (CFRAM) segments. In light of the SEM examination of the tried parts, connections between aftereffects of the mechanical test and its micro structural parts were examined. CFRAM parts are lightweight yet solid materials with a wide scope of expected applications in the vehicle industry, aviation, sports products, and clinical apparatuses.

CFRAM parts work with both state-of-the-art 3D printing innovation and fiber support to inprove the mechanical properties. Created parts are lightweight compared to metals, have solid mechanical properties, and short assembling time. What's more, the thermoplastic polymer utilized for CFRAM segments makes items recyclable. Light weight, low cost, and adequate thermomechanical properties make fiber-reinforced 3D-printed polymer composites incredible up-and-comers that can substitute for metals in a wide scope of uses. Nanomaterials, miniature particles, and short and long filaments have been added to polymers to improve mechanical properties [50].

Developing natural products related to the creation, removal, and reusing of manufactured fiber-based polymer composites has triggered the advancement of eco-friendly composites for different applications like autos, marine uses, compounds, infrastructure, outdoor supplies, etc. Along with natural strands like kenaf, jute, oil palm, cotton, flax, banana, and hemp, sisal is gaining favor as it is easily accessible, less expensive, eco-accommodating, as well as possessing similar mechanical properties to hemp, banana, and jute. Sisal fiber will assume a vital role in the manufacture of a differing scope of underlying and non-primary mechanical items with various polymer matrices.

Figure 3.5 Applications of polymer composites.

Cement has been utilized for numerous underlying applications for many years. But because of the expanding cost of the material, there is a growing need for alternate materials or a way to lower its cost. Utilization of sisal-concrete composites through material tiles, dividing sheets, level sheets, and layered sheets, and so on, have for some time been explored in non-industrial nations for the development of low-cost homes and structures. Among different plant strands, sisal filaments are favored in the field of development industry because of their great warm and acoustic properties, along with their incredible rigidity and strength [51]. Figure 3.5 represents applications of polymer composites.

3.6 CONCLUSION

In the day-by-day discovery of innovative environmentally sustainable components in the world of technology products, some new products have come to the fore for utilization. The only constraint in the usage of certain components is the weight of the product to be utilized. Polymer materials have been one solution for this problem for the past two decades. Introduction of many combinations of composite materials with different properties allows utilization for various desired applications.

This chapter discussed the role of natural, synthetic, particulate, fillers, and application of polymer materials for lightweight applications. The conclusion drawn through this research is discussed below:

✓ In polymer composites natural fibers are used as reinforcement materials. These composites proved to be very successful. Owing to their origin, they can behave based on the chosen applications. Major areas such as marine, mechanical, aerospace, and automobile engineering are correctly utilizing the functional approaches of NFRC for various lightweight applications.

✓ The development of synthetic fibers are considered as unpredictable surface morphology, which upgrades interlocking at the fiber-composite interface and subsequently working on mechanical properties of the composites. In this, it is worth noting that the surface state of a hydrophilic normal fiber plays a significant part in its anything but a hydrophobic polymer framework.

✓ The successful expansion of thermoplastics as a lightweight material for auto applications and other business purposes prompts further study of the accessible polymer assets. In this examination, the idea of improving the exhibition of reused squander plastics as polymer-based composites was inspected. Aggregation of filler particles during composite creation affects their non-homogenous dissemination and preparation issues and causes imperfect surface quality and voids that decay mechanical properties.

✓ The structure, properties, and quantity of layers of the graphene-based materials are closely identified with the strategy picked for their creation. These cycles can be isolated into two principal classifications: the hierarchical and granular perspectives.

✓ The CPCs can be used as antistatic or anticorrosive materials, electromagnetic interference (EMI) shielding materials, and conductors. Interestingly, it was found that the conductive network in a specific polymer matrix had a response to the external environments, such as mechanical stress, temperature, solvent, vapor, and so on. The areas of MEMS, drugs, sensors, tissue engineering, intelligent application system, hulls, and wings of the airplane and marine applications were affected as well.

REFERENCES

1. Nurazzi, N. M., M. R. M. Asyraf, A. Khalina, N. Abdullah, H. A. Aisyah, S. Ayu Rafiqah, and F. A. Sabaruddin. (2021). A review on natural fiber reinforced polymer composite for bullet proof and ballistic applications. *Polymers*, 13(4), 646.
2. Suresh, S., D. Sudhakara, and B. Vinod. (2020). Investigation on industrial waste eco-friendly natural fiber-reinforced polymer composites. *Journal of Bio- and Tribo-Corrosion*, 6(2), 1–14.

3. Tavares, Tânia D., Joana C. Antunes, Fernando Ferreira, and Helena P. Felgueiras. (2020). Biofunctionalization of natural fiber-reinforced biocomposites for biomedical applications. *Biomolecules*, 10(1), 148.

4. Jeyapragash, R., V. Srinivasan, and S. J. M. T. P. Sathiyamurthy. (2020). Mechanical properties of natural fiber/particulate reinforced epoxy composites–A review of the literature. *Materials Today: Proceedings*, 22, 223–1227.

5. Zaini, E. S., M. D. Azaman, M. S. Jamali, and K. A. Ismail. (2020). Synthesis and characterization of natural fiber reinforced polymer composites as core for honeycomb core structure: A review. *Journal of Sandwich Structures & Materials*, 22(3), 525–550.

6. Joseph, Jomy, Prithvi Raj Munda, Manoj Kumar, Ajay M. Sidpara, and Jinu Paul. (2020). Sustainable conducting polymer composites: Study of mechanical and tribological properties of natural fiber reinforced PVA composites with carbon nanofillers. *Polymer-Plastics Technology and Materials*, 59(10), 1088–1099.

7. Chegdani, Faissal, Behrouz Takabi, Mohamed El Mansori, Bruce L. Tai, and Satish TS Bukkapatnam. (2020). Effect of flax fiber orientation on machining behavior and surface finish of natural fiber reinforced polymer composites. *Journal of Manufacturing Processes*, 54, 337–346.

8. Sumesh, K. R., K. Kanthavel, and V. Kavimani. (2020). Peanut oil cake-derived cellulose fiber: Extraction, application of mechanical and thermal properties in pineapple/flax natural fiber composites. *International Journal of Biological Macromolecules*, 150, 775–785.

9. Maran, M., R. Kumar, P. Senthamaraikannan, S. S. Saravanakumar, S. Nagarajan, M. R. Sanjay, and Suchart Siengchin. (2020). Suitability evaluation of sida mysorensis plant fiber as reinforcement in polymer composite. *Journal of Natural Fibers*, 5, 1–11.

10. Hassan, Tufail, Hafsa Jamshaid, Rajesh Mishra, Muhammad Qamar Khan, Michal Petru, Jan Novak, Rostislav Choteborsky, and Monika Hromasova. (2020). Acoustic, mechanical and thermal properties of green composites reinforced with natural fibers waste. *Polymers*, 12(3), 654.

11. Kumar, Santosh, Divya Zindani, and Sumit Bhowmik. (2020). Investigation of mechanical and viscoelastic properties of flax-and ramie-reinforced green composites for orthopedic implants. *Journal of Materials Engineering and Performance*, 29, 3161–3171.

12. Binoj, J. S., R. Edwin Raj, Shukur Abu Hassan, M. Mariatti, Suchart Siengchin, and M. R. Sanjay. (2020). Characterization of discarded fruit waste as substitute for harmful synthetic fiber-reinforced polymer composites. *Journal of Materials Science* 55(20), 8513–8525.

13. Lokesh, P., TSA Surya Kumari, R. Gopi, and Ganesh Babu Loganathan. (2020). A study on mechanical properties of bamboo fiber reinforced polymer composite. *Materials Today Proceedings*, 22, 897–903.

14. Mohan, Velram Balaji, and Debes Bhattacharyya. (2020). Mechanical characterization of functional graphene nanoplatelets coated natural and synthetic fiber yarns using polymeric binders. *International Journal of Smart and Nano Materials*, 11(1), (2020): 78–91.

15. Oliveira, Michelle Souza, Fernanda Santos da Luz, Andressa Teixeira Souza, Luana Cristyne da Cruz Demosthenes, Artur Camposo Pereira, Fábio de Oliveira Braga, André Ben-Hur da Silva Figueiredo, and Sergio Neves

Monteiro. (2020). Tucum fiber from amazon Astrocaryum vulgare palm tree: Novel reinforcement for polymer composites. *Polymers*, 12(10), 2259.

16. Hassan, Mohamad Zaki, S. M. Sapuan, Zainudin A. Rasid, Ariff Farhan Mohd Nor, Rozzeta Dolah, and Mohd Yusof Md Daud. (2020). Impact damage resistance and post-impact tolerance of optimum banana-pseudo-stem-fiber-reinforced epoxy sandwich structures. *Applied Sciences*, 10(1),684.

17. Asim, Mohammad, Mohd T. Paridah, M. Chandrasekar, Rao M. Shahroze, Mohammad Jawaid, Mohammed Nasir, and Ramengmawii Siakeng. (2020). Thermal stability of natural fibers and their polymer composites. *Iranian Polymer Journal*, 29, 625–648.

18. Bambach, Mike R. (2020). Direct comparison of the structural compression characteristics of natural and synthetic fiber-epoxy composites: Flax, jute, hemp, glass and carbon fibers. *Fibers*, 8(10), 62.

19. Yuhazri, M. Y., A. J. Zulfikar, and A. Ginting. (2020). Fiber reinforced polymer composite as a strengthening of concrete structures: A review. *In IOP Conference Series: Materials Science and Engineering*, 1, 012135.

20. Chavhan, Ganesh R., and Lalit N. Wankhade. (2020). Improvement of the mechanical properties of hybrid composites prepared by fibers, fiber-metals, and nano-filler particles–A review. *Materials Today: Proceedings*, 27, 72–82.

21. Shukla, Bishnu Kant, Anit Raj Bhowmik, and Pushpendra Kumar. (2020). Use of waste synthetic fiber reinforcement in environmental friendly and economic pavement construction, *Journal of Green Engineering*, 10(1), 1–14.

22. Ramesh, V., and P. Anand. (2021). Evaluation of mechanical properties on Kevlar/Basalt fiber reinforced hybrid composites. *Materials Today: Proceedings*, 39, 1494–1496.

23. Sharma, Aanchna, S. Anand Kumar, and Vinod Kushvaha. (2020). Effect of aspect ratio on dynamic fracture toughness of particulate polymer composite using artificial neural network. *Engineering Fracture Mechanics*, 228, 106907.

24. Khotbehsara, Mojdeh Mehrinejad, Allan Manalo, Thiru Aravinthan, Joanna Turner, Wahid Ferdous, and Gangarao Hota. (2020). Effects of ultraviolet solar radiation on the properties of particulate-filled epoxy based polymer coating. *Polymer Degradation and Stability*, 181, 109352.

25. Oladele, I. O., A. A. Adediran, A. D. Akinwekomi, Miracle Hope Adegun, O. O. Olumakinde, and O. O. Daramola. (2020). Development of ecofriendly snail shell particulate-reinforced recycled waste plastic composites for automobile application. *The Scientific World Journal*, 12 (3), 262–286.

26. Samal, Sneha. (2020). Effect of shape and size of filler particle on the aggregation and sedimentation behavior of the polymer composite. *Powder Technology*, 366, 43–51.

27. Jagadeesh, Praveenkumara, Yashas Gowda Thyavihalli Girijappa, Madhu Puttegowda, Sanjay Mavinkere Rangappa, and Suchart Siengchin. (2020). Effect of natural filler materials on fiber reinforced hybrid polymer composites: An overview. *Journal of Natural Fibers*, 12(6), 1–16.

28. Presto, Dillon, John Meyerhofer, Grant Kippenbrock, Suresh Narayanan, Jan Ilavsky, Sergio Moctezuma, Mark Sutton, and Mark D. Foster. (2020). Influence of silane coupling agents on filler network structure and stress-induced particle rearrangement in elastomer nanocomposites. *ACS Applied Materials & Interfaces*, 12 (42), 47891–47901.

29. Khan, Mohammad Zahid Rayaz, Sunil Kumar Srivastava, and M. K. Gupta. (2020). A state-of-the-art review on particulate wood polymer composites: Processing, properties and applications. *Polymer Testing*, 5(16), 106721.

30. Ahmed, Waleed, Sidra Siraj, and Ali H. Al-Marzouqi. (2020).3D printing PLA waste to produce ceramic based particulate reinforced composite using abundant silica-sand: Mechanical properties characterization. *Polymers*, 12(11), 2579.

31. Stejskal, Jaroslav. (2020). Conducting polymers are not just conducting: A perspective for emerging technology. *Polymer International*, 69(8),662–664.

32. Donato, Ricardo K., and Alice Mija. (2020). Keratin associations with synthetic, biosynthetic and natural polymers: An extensive review. *Polymers*, 12(1), 32.

33. Tian, Bingren, and Jiayue Liu. (2020). The classification and application of cyclodextrin polymers: A review. *New Journal of Chemistry*, 44 (22), 9137–9148.

34. Alsubari, S., M. Y. M. Zuhri, S. M. Sapuan, M. R. Ishak, R. A. Ilyas, and M. R. M. Asyraf. (2021). Potential of natural fiber reinforced polymer composites in sandwich structures: A review on its mechanical properties. *Polymers*, 13 (3), 423.

35. da Luz, Fernanda Santos, Fabio da Costa Garcia Filho, Maria Teresa Gomez Del-Rio, Lucio Fabio Cassiano Nascimento, Wagner Anacleto Pinheiro, and Sergio Neves Monteiro. (2020). Graphene-incorporated natural fiber polymer composites: A first overview. *Polymers*, 12(7), 1601.

36. van Grootel, Alexander, Jiyoun Chang, Brian L. Wardle, and Elsa Olivetti. (2020). Manufacturing variability drives significant environmental and economic impact: The case of carbon fiber reinforced polymer composites in the aerospace industry. *Journal of Cleaner Production*, 261, 121087.

37. Pastore, R., A. Delfini, M. Albano, A. Vricella, M. Marchetti, F. Santoni, and F. Piergentili. (2020). Outgassing effect in polymeric composites exposed to space environment thermal-vacuum conditions. *Acta Astronautica*, 170, 466–471.

38. Kenned, Jack J., K. Sankaranarayanasamy, J. S. Binoj, and Suresh Kumar Chelliah. (2020). Thermo-mechanical and morphological characterization of needle punched non-woven banana fiber reinforced polymer composites. *Composites Science and Technology*, 185, 107890.

39. Ray, Krutibash, Hemalata Patra, Anup Kumar Swain, Bibhudatta Parida, Sourabh Mahapatra, Asit Sahu, and Suryakanta Rana. (2020). Glass/jute/sisal fiber reinforced hybrid polypropylene polymer composites: Fabrication and analysis of mechanical and water absorption properties. *Materials Today: Proceedings*, 33, 5273–5278.

40. Meng, Jinxin, Yong Wang, Haiyang Yang, Panding Wang, Qin Lei, Hanqiao Shi, Hongshuai Lei, and Daining Fang. (2020). Mechanical properties and internal microdefects evolution of carbon fiber reinforced polymer composites: Cryogenic temperature and thermocycling effects. *Composites Science and Technology*, 191, 108083.

41. Rahman, Ateeq, Ilias Ali, Saeed M. Al Zahrani, and Rabeh H. Eleithy. (2011). A review of the applications of nanocarbon polymer composites. *Nano*, 6(3), 185–203.

42. George, Suchi Mercy, and Balasubramanian Kandasubramanian. (2020). Advancements in MXene-Polymer composites for various biomedical applications. *Ceramics International*, 46(7), 8522–8535.

43. Tanasa, Fulga, Madalina Zanoaga, Carmen-Alice Teaca, Marioara Nechifor, and Asim Shahzad. (2020). Modified hemp fibers intended for fiber-reinforced polymer composites used in structural applications—A review. I. Methods of Modification. *Polymer Composites*, 41(1), 5–31.
44. Correia, Daniela Maria, Liliana Correia Fernandes, Pedro Manuel Martins, Clara García-Astrain, Carlos Miguel Costa, Javier Reguera, and Senentxu Lanceros-Méndez. (2020). Ionic liquid–polymer composites: A new platform for multifunctional applications. *Advanced Functional Materials*, 30(24), 1909736.
45. Johnson, R., V. Arumuga Prabu, P. Amuthakkannan, and K. Arun Prasath. A review on biocomposites and bioresin based composites for potential industrial applications. (2017). *Reviews on Advanced Materials Science*, 49(1), 22–30.
46. Prasath, K. Arun, P. Amuthakkannan, V. Manikandan, R. Jegadeesan, and M. Selwin. (2019). Novel topological approach in mechanical properties of basalt/flax hybrid composites. *In AIP Conference Proceedings*, 2128(1), 020001.
47. Chen, Jianwen, Yutian Zhu, Jinrui Huang, Jiaoxia Zhang, Duo Pan, Juying Zhou, Jong E. Ryu, Ahmad Umar, and Zhanhu Guo. (2021). Advances in responsively conductive polymer composites and sensing applications. *Polymer Reviews*, 61(1), 157–193.
48. Bayram, Yakup, Yijun Zhou, Bong Sup Shim, Shimei Xu, Jian Zhu, Nick A. Kotov, and John L. Volakis. (2010). E-textile conductors and polymer composites for conformal lightweight antennas. *IEEE Transactions on Antennas and Propagation*, 58 (8), 2732–2736.
49. Rajapaksha, L. D., H. A. D. Saumyadi, A. M. P. B. Samarasekara, D. A. S. Amarasinghe, and L. Karunanayake. (2017). Development of cellulose based light weight polymer composites. In *2017 Moratuwa Engineering Research Conference (MERCon)*, Moratuwa, Srilanka, 182–186.
50. Mohammadizadeh, M., A. Imeri, I. Fidan, and M. Elkelany. (2019). 3D printed fiber reinforced polymer composites-Structural analysis. *Composites Part B: Engineering*, 175, 107112.
51. Senthilkumar, K., N. Saba, N. Rajini, M. Chandrasekar, M. Jawaid, Suchart Siengchin, and Othman Y. Alotman. (2018). Mechanical properties evaluation of sisal fibre reinforced polymer composites: A review. *Construction and Building Materials*, 174, 713–729.

Chapter 4

Chemical property and characteristics of polymer

A. Sofi, Joshua Jeffrey, and Abhimanyu Singh Rathor
Vellore Institute of Technology, Vellore, India

CONTENTS

DOI: 10.1201/9781003252108-4

4.1 INTRODUCTION

Materials in growing industries are always in need of improvement—to improve strength, wear resistance, stiffness, density, and to lower costs with improved sustainability. Polymers enhance the performance and versatility of materials while keeping the material lightweight. Fiber-reinforced polymers (FRP) are composite material that comprises polymer matrix reinforced with fiber. Composites made of fiber-reinforced polymers are even being investigated as a replacement for standard materials like concrete and steel.

Although structural fiber composites are rarely used in applications with high load-bearing capacity requirements, repairing old buildings is one of their most prevalent uses in the construction sector. When it comes to structural applications, polymer composites consist of two or more engineered or natural materials with a diverse spectrum of physical or chemical qualities that remain separate and autonomous within the final structure [1]. Some studies have also shown that in reinforced and stressed concrete, polymer composite materials are also used to replace steel [2].

In addition to being lightweight, noncorrosive, simple to construct, tailorable to performance needed, as well as exhibiting high performance characteristics, FRP composites are also easy to construct. Due to these characteristics, fiber-reinforced polymer composites have found a significant number of applications in various fields—and the uses are growing. Thermoplastic polymers and thermosetting polymers are the two most common forms of polymers. As thermoplastic matrix materials are made up of one-dimensional or two-dimensional molecular structures, they tend to soften at higher temperatures and then roll back their properties as they cool.

Thermoset polymers, on the other hand, are strongly cross-linked polymers that cure with heat alone—or with heat and pressure and/or light irradiation. Thermoset polymers have good properties because of this structure, such as tremendous flexibility for modifying desired ultimate characteristics, high strength, and modulus [3]. Epoxy, polyester, polyurethane, and phenolic resins are common thermosetting resins with enhanced performance requirements. Fiber-reinforced concrete (FRC) is divided into four groups of fiber material according to American Concrete Institute Committee 544. Steel fiber is referred to as SFRC; glass fiber FRC is referred to as GFRC; synthetic fiber FRC, including carbon fibers, is referred to as SNFRC; and natural fiber FRC is referred to as NFRC [4].

4.2 NATURAL FIBER AS REINFORCEMENT IN POLYMERS

Natural fibers are once again at the forefront because of the necessity to turn to alternatives with a low environmental footprint. When using natural fibers, some factors must be considered, such as the fiber's type, age, shape, harvesting method, processing, and ability to absorb water. Furthermore, their chemical composition influences their behavior, and the uniformity of cellulose and lignin affects the hydration process and, ultimately, strength growth [5].

Fibers are mainly used to reduce plastic shrinkage and drying shrinkage, as well as to increase tensile strength, toughness, and durability. Their functioning is determined by their geometry such as length, diameter, shape, and orientation, once dispersed in the mixture [6]. Because of the advantageous properties and better benefits of natural fiber over synthetic fiber—as far as its somewhat low weight, minimal expense, less harm to handling hardware and processing equipment, great relative mechanical properties like tensile modulus and flexural modulus, and improved finish on surface completion of formed parts composite—fiber-reinforced polymer matrix boasts significant consideration in various applications [7].

Table 4.1 shows the chemical composition of some natural fibers. The flexibility in processing, biodegradability, and low health risks are all advantages. By integrating a resilient, durable, and lightweight natural fiber with a polymer, NFPCs (natural fiber polymer composites) with high specific stiffness and strength can be developed (thermoplastic and thermoset) [8]. Some of the most well-known types are hemp, jute, coconut, bamboo, sisal, rice husk, palm, flax, cotton, and sugarcane [9].

Composite materials constructed of high-strength natural fibers incorporated in a polymer matrix such as jute, oil palm, sisal, kenaf, and flax are

Table 4.1 Chemical composition of some renowned natural fibers [7]

Fiber	Cellulose (wt%)	Hemicellulose (wt%)	Lignin (wt%)	Waxes (wt%)
Bagasse	55.2	16.8	25.3	–
Bamboo	26–43	30	21–31	–
Flax	71	18.6–20.6	2.2	1.5
Kenaf	72	20.3	9	–
Jute	61–71	14–20	12–13	0.5
Hemp	68	15	10	0.8
Ramie	68.6–76.2	13–16	0.6–0.7	0.3
Abaca	56–63	20–25	7–9	3
Sisal	65	12	9.9	2
Coir	32–43	0.15–0.25	40–45	
Oil palm	65	–	29	–
Pineapple	81	–	12.7	–
Curaua	73.6	9.9	7.5	–
Wheat straw	38–45	15–31	12–20	–
Rice husk	35–45	19–25	20	–
Rice straw	41–57	33	8–19	8–38

known as natural fiber polymer composites (NFPC) [10]. To get good NFPC characteristics, considerable fiber loading is usually required.

4.2.1 Treatments of natural fibers

Natural fiber treatments are necessary to alter fiber surface properties to enhance adhesion to a variety of matrices and materials such as cement mortar, wood, and plastics [11]. The fiber treatment techniques may be divided into two categories: firstly, chemical treatments (to increase the interfacial characteristics of the fiber matrix and the fiber durability of cement composites); and secondly, physical treatments (to improve natural fiber attributes such as tensile strength, elasticity, and extensibility) [12]. Table 4.2 shows the physical and mechanical properties of natural fibers.

Physical treatments alter the structural and surface morphology of polymers, affecting their mechanical interaction. These treatments include stretching, calendar therapy, corona and thermal therapy, plasma therapies, yarn manufacturing, etc. Physical treatment has little effect on the chemical structure of fibers. This indicates that in most applications a stronger mechanical interaction between the fiber and the matrix enhances the interface [14]. Natural fibers are woven or spun into threads, which are then refined into preforms for yarning. Yarn twist, on the other hand, is primarily accountable for the occurrence of the fiber to be bound to a stronger yarn by friction in most short fiber yarns. As a response, rotating the fibers is

Table 4.2 Physical and mechanical properties of natural fibers [13]

Fiber	Density (g/cm³)	Tensile strength (MPa)	Young's modulus (GPa)	Elongation at break (%)
OPEFB	0.7–1.55	248	3.2	2.5
Flax	1.4	88–1500	60–80	1.2–1.6
Hemp	1.48	550–900	70	1.6
Jute	1.46	400–800	10–30	1.8
Ramie	1.5	500	44	2
Coir	1.25	220	6	15–25
Sisal	1.33	600–700	38	2–3
Abaca	1.5	980	–	–
Cotton	1.51	400	12	3–10
Kenaf (bast)	1.2	295	–	2.7–6.9
Kenaf (core)	0.21	–	–	–
Bagasse	1.2	20–290	19.7–27.1	1.1
Henequen	1.4	430–580	–	3–4.7
Pineapple	1.5	170–1672	82	1–3
Banana	1.35	355	33.8	53

required to provide a minimal coherence between them; otherwise, producing a short fiber yarn with significant tensile strength is impossible [15].

Bleaching, vinyl grafting, acetylating, benzoylation, alkaline treatment, peroxide treatment, and therapy of various coupling agents are some chemical processing treatments that cater to the extent of interfacial bonding between the natural fiber and the polymer matrix. When comparing the chemical treatments listed, alkali treatment is the most prevalent [16]. It's a low-cost, very efficient surface treatment for natural fibers that enhances their mechanical properties. Alagirusamy et. al. reported that alkali treatment removes a portion of the cell wall's external surface, which includes hemicellulose, lignin, wax, and oils. This results in rougher surfaces for improved fiber interlocking, matrix penetration, and fiber-matrix contact area [17].

4.2.2 Coconut, kelp, and jute fibers in mortars

Kesikidou et. al. investigated the following three types of natural fibers (jute, coconuts, and kelp). Jute is a bast fiber used as a backing material for tufted carpets in the form of burlap, sacking, and twine. The plant components lignin, cellulose, and pectin make up a large portion of the jute fiber [18]. Kelp is a large brown seaweed that grows at coastal fronts all around the world in shallow, nutrient-rich seawater. Its use in cement and lime mortars was investigated and compared with each other and a control sample in order to ascertain their application based on origins and physical characteristics. The fibers were cut by hand with a length of approximately 1 cm, and they were added in a 1.5% w/v of the mortar [19].

From Table 4.3, an inference can be drawn that jute fibers have more affinity for water absorption, whilst coconut fibers absorb the least amount of water. Furthermore, kelp fibers present a high amount of soluble content in when compared to coconut fibers, which has the lowest water-soluble content. This information must be taken into account when it comes to mix design and calculating the appropriate water-cement ratios. Prismatic specimens (40×40×160) mm using cement type OPC 53 grade and lime mortars were produced and tested, and all the sample specimens were subjected to flexural and compressive testing at 28 days with a universal testing machine (UTM). Because of the higher porosity after evaporation in cement mortar

Table 4.3 Fiber characteristics [19]

Fiber type	Cellulose (%)	Lignin (%)	Water absorption (%)	Water soluble content (%)	Width (average) (mm)
Jute	72	13	84	6.0	0.08
Coconut	43	45	73	5.5	0.19
Kelp	61.8	29.8	80	18.0	3.26

with coconut and kelp fibers observed from the extra water added due to the fiber's high water absorption values, when compared to a conventional control mortar mix, the compression tests indicated a drop in strength from 3% for coconut fibers to 6.8% for kelp fibers. From that, it may be inferred that natural fibers of jute, coconut, and kelp work better with lime in terms of strength growth, whereas these fibers in cement behave very differently. Kelp fibers showed the highest flexural strength and C-R fibers attained its peak compressive strength at 62 MPa. According to research, the chemical consistency of lignin, cellulose, and hemicellulose fibers may have a detrimental impact on cement hydration due to their disintegration in water [20].

4.2.3 Sugarcane bagasse

It is well known that softwood pulp fibers can be used as a substitute for asbestos fibers in the Hatscheck manufacturing process for fiber-reinforced cement products [21]. Bagasse is the lignocellulosic residue left after the juice from the sugar cane stalk has been extracted. Bagasse sugarcane is a waste product of the sugar industry, currently used as a fuel in sugar mill boilers [22]. Bilba et. al. concluded that mixing raw unrefined whole bagasse with conventional cement types has the effect of delaying the setting time and lowering the maximum hydration temperature of the setting [20]. By reducing the heat of hydration for the setting of cement, the concrete is less prone to cracking by plastic shrinkage. Through experimental methods, Riza Wirzwan et. al. inferred that the tensile strength of sugarcane bagasse was typically 170–290 MPa, while the modulus of elasticity was 15–19 GPa on average [23]. In further studies conducted, Young's modulus values peaking at 27.1 GPa were reported, together with the ultimate tensile strength of 222 MPa and an elongation at a break of 1.1% were also inferred in studies conducted [24]. By providing a suitable compromise between setting time and hydration temperature, this polymer has considerable appeal for building materials.

4.3 SYNTHETIC FIBER AS REINFORCEMENT IN POLYMERS

Synthetic fibers have seen rapid application and development in recent years and have proven quite useful in the construction industry. Despite the fact that there are certain drawbacks in mechanical performance, the higher durability and physical properties of synthetic fibers, along with effective and relatively inexpensive reinforcement, have increased their use in concrete. It has become an alternative to asbestos, steel, and glass fibers. Synthetic fibers used in cement matrices include polypropylene (PP), polyvinyl alcohol (PVA), polyethylene (PE), acrylics (PAN), carbon reinforcements, polyester (PES), polyamides (PA), and aramid. Table 4.4 shows the physical properties of various synthetic fibers.

Table 4.4 Physical properties of some polymeric fibers [25]

Fiber type	Specific gravity	Tensile strength (MPa)	Elastic modulus (GPa)	Ultimate elongation (%)
Acrylic	1.17	207–1000	14.6–19.6	7.5–50.0
Aramid I	1.44	3620	62	4.4
Aramid II (high modulus)	1.44	3620	117	2.5
Nylon	1.16	965	5.17	20.0
Polyester	1.34–1.39	896–1100	17.5	
Polyethylene	0.96	200–300	5.0	3.0
Polypropylene	0.90–0.91	310–760	3.5–4.9	15.0

The synthetic fibers are made from both naturally occurring and synthesized macromolecules. They have a flexible, macroscopically homogenous body with a high aspect ratio and low cross-sectional area. The chemical, physical, and mechanical properties of these fibers are largely governed by the three-dimensional arrangement of these polymer strands [25].

4.3.1 Polypropylene fiber polymer (PP)

Polypropylene polymer is a linear hydrocarbon with some properties similar to polyethylene. However, the methyl side molecules connected to alternate carbon atoms in the polymer chain alter certain properties, producing stiffness, raising the melting temperature, and making the polymer comparatively less stable than PE with respect to oxidation. PP's microstructure is designed to promote crystal formation; as a result, the chemical resistance and thermal stability are both outstanding.

PP fibers offer many characteristics that make them ideal for use in concrete matrices. In the alkaline environment of concrete, they are chemically inert and very stable. With the low cost of raw materials, they additionally have a substantially high melting point. The polymer's surface is hydrophobic, which implies that it won't absorb water. PP fibers are known to form a weak interconnection with the concrete mix and hence additional processes have to be performed to ensure effective bonding.

The rheological parameters of fresh concrete are significantly influenced by PP fibers. The slump range of PP fiber-reinforced concrete (FRC) drops as volume concentration of the fiber rises, as per workability tests. The inclusion of PP fiber at 0.5% by volume reduced the slump of a 0.5 water-cement ratio mix from 89 mm to 13 mm, according to Al-Tayyib et. al., even though the mix flowed well and responded well to vibration. Water-reducing admixtures or increasing the water quantity are solutions that can compensate for the slump reduction [26].

4.3.2 Polyethylene fiber polymer (PE)

Yang et. al. reported in their findings that high-strength PE fiber has a low melting point of 80°C to 147°C and outstanding specific strength and modulus [27]. Polyethylene fiber has the advantage of being able to be manufactured with a reasonably high modulus of elasticity. Polyethylene fibers with a high modulus have a tensile strength of around 2.6 GPa, Young's modulus of elasticity of around 117 GPa, and elongation at break of 5 to 8%, with specific gravity of 0.97. When used in cement paste, high-modulus PE fiber improves the material's flexural strength, toughness, and impact resistance [28]. Soroushian et al. indicated that PE fibers used at 0.1% by volume in concrete increased average flexural and impact strengths by 21% and 5.8 times, respectively [29].

4.3.3 Acrylics (PAN)

Acrylic fibers are classified as fibers that contain at least 85% acrylonitrile (AN) by weight. West Germany recently produced a unique acrylic fiber with long polymer chains made from a high-performance polyacrylonitrile raw material. The fiber's tensile strength and elastic modulus, respectively, can reach 900–1000 MPa and 17–19.5 GPa. The acrylic fibers have high tensile strength and elastic modulus, as well as superior acid and alkali resistance and a low cost. These fibers can improve the toughness of mortar and concrete, as well as minimize fissure formation and microcracks. Fiber-matrix adhesion would most likely be improved by uniform hydrolysis of the fiber. However, excessive hydrolysis would degrade the fibers' mechanical characteristics, preventing their application [27]. According to the ring test reported by Hahne et al., adding 2.5% acrylic fiber to reinforced concrete reduces shrinkage cracking caused by setting and hardening.

4.3.4 Carbon fibers

Carbon fibers are known to be chemically stable in alkaline conditions, to resist abrasions, and are stable at high temperatures. They are also medically acceptable, have proved to be as strong as steel, and are more chemically stable when compared to glass fibers. Carbon fibers are made up of tows, with each consisting of multiple filaments. The filaments have a diameter of 7–15 μm and are made up of tiny crystallites of "turbostratic" graphite [25]. Their density is lower when compared to steel fibers, and they have one of the highest strength-to-density ratios of all the fiber types. Their drawback is high cost, which impedes their use on an industrial scale for Portland cement applications. The properties of carbon fibers are shown in Table 4.5.

The two major methods for producing carbon fibers are dependent on their source materials: acrylic carbon fibers or petroleum and coal tar pitch. Due to the high cost of PAN-based carbon fibers, their compositions have not been encouraged for use in the field. Pitch-based fibers have a much

Table 4.5 Properties of carbon fibers [30]

Property	PAN		Pitch
	Type I	Type II	
Diameter (μm)	7.0–9.7	7.6–8.6	18
Density (kg m⁻³)	1950	1750	1600
Modulus of elasticity (GPa)	390	250	30–32
Tensile strength (MPa)	2200	2700	600–750
Elongation at break (%)	0.5	1.0	2.0–2.4
Coefficient of thermal expansion ($\times 10^{-6}$ °C⁻¹)	−0.5 to −1.2 (parallel) 7–12 (radial)	−0.1 to −0.5 (parallel) 7–12 (radial)	–

lower elasticity and strength moduli than PAN carbon fibers, but they are also less expensive. Pitch-based carbon fibers retain superior characteristics over many other synthetic fibers, making them appropriate for use in cement and concrete applications [25].

4.3.5 Polyesters (PES)

In the creation of this fiber, molten polymer is extruded and then carried into the air at ambient room temperatures through spinnerets. As a result, the fibers cool quickly and become amorphous and feeble. Changes in manufacturing procedures can significantly alter the physical and chemical properties of polyester fibers [25]. Patel et al. examined the attributes of polyester FRC and found that adding polyester fiber at 1.0% (by volume) increased impact strength by 75%, split tensile strength by 9%, flexural strength by 7%, and compressive strength by 5% [31]. The modulus of elasticity and shear strength values remained unchanged. Wang et. al. investigated polyester fibers in Portland cement concrete and found that there is a rapid decline in strength in the cement matrix [25]. Because of their chemical deterioration, the American Concrete Institute Committee decided that polyester fibers were inefficient as reinforcing materials that can be used in PPC (Portland cement) products [32].

4.3.6 Aramid fibers

Aramids are a polymer product with market names such as Kevlar, Nomex, Technorac, and Twaron Fibers. Kevlar is the material used to manufacture bulletproof body armor for military and law-enforcement agencies worldwide. The aromatic rings give the polymer rigidity, and these stiff chains are typically parallelly oriented to the axis of the fiber in the manufacturing of these fibers, resulting in a modulus of elasticity of up to 130 GPa. They are utilized in both rigid and flexible systems as structural reinforcement while also providing friction, abrasion, and ballistic resistance [27].

In the findings from the study conducted by A. Nanni, synthetic fibers formed by cutting an epoxy-impregnated, braided bundle of aramid filaments behave similarly to steel fibers when used as reinforcement in concrete and slurry matrices [33]. The lack of corrosion concerns and superior performance of epoxy-coated aramid over polypropylene are two advantages of epoxy-coated aramid over steel. Commercially produced Kevlar-49 fibers have also been utilized to strengthen cement free of macro-defects, primarily to improve toughness and impact performance [25].

4.3.7 Polyvinyl alcohol (PVA) fibers

Kuralon and Vinylon are brand names for PVA fibers. The advantages of using fibers to make fiber-reinforced concrete products are as follows: high aspect ratio, high ultimate tensile strength, strong chemical compatibility with Portland cement, good affinity for water, faster drainage rate, and no health risks associated with their use [25]. Unitika Kasei Ltd, Japan, invented high-modulus Vinylon fiber, which has been hailed as one of the most promising new materials for replacing asbestos in fiber-reinforced cementitious composite materials. It possesses numerous appealing properties, including increased strength (1.23 GPa), high modulus (2.95 GPa), good alkali resistance, effective cement bonds, and a specific gravity of approximately 1.3.

4.3.8 Polyamide fibers (PA)

Polyamides (PA) are commonly referred to as nylons. The commercial success of nylon 6 and nylon 6,6 is attributed to the polymers' excellent characteristics and a cost-effective raw material base. Polyamide fibers are sold in single-filament form and come in a variety of lengths. Because these fibers are so thin, their number per weight (fiber count) with a fiber length of 19 mm is in the region of 77 million per kg [25]. Walton and Majumdar found that tiny concentrations of these fibers significantly improve the composite's impact resistance without affecting its tensile or flexural strength [25, 34]. The high-impact resistance of the nylon-fiber composite is due to the stretching and pulling out of the fibers that happens at large stresses after the matrix fails and at a lower load. If composites could be constructed to support rising loads after the matrix cracks, that would be ideal. After matrix failure, this can be accomplished by enhancing the stress transfer to the fibers from the matrix after its failure.

4.4 HYBRID FIBERS

When compared to thermoset composites, the thermoplastic composites that have been reinforced with natural fibers exhibit poor performance of compressive and flexural strength. Hence a blend of organic and inorganic fibers can be used to make cement-based composites. These hybrid fiber cement systems are made up of two or more types of fiber. The goal is to maximize the desired composite qualities by utilizing the best properties of the individual fibers. In typical use, the high-impact strength produced from the synthetic fibers like nylon and PP will be stable for a lengthy period [34]. The inclusion of a second fiber, like asbestos or glass, can increase the tensile properties and flexural characteristics of these composites. When a filler content of 25% hemp and 15% glass was included in a composite construction by weight, the flexural strength was 101 MPa and flexural modulus was found to be 5.5 GPa [35].

SEM analysis of an oil palm/kenaf fiber-reinforced epoxy hybrid composite exhibited strong interfacial bonding between the fiber and the matrix, indicating improved tensile and flexural capabilities. Furthermore, in comparison with conventional composites, the oil palm hybrid composite absorbed more energy throughout the impact loading process, making the hybrid material a viable contender in the automobile industry [36]. The hybridization of fibers is a convincing method in which multiple fiber types are mixed in a composite matrix to reduce the disadvantages of one type of fiber while maintaining the beneficial qualities of another.

4.5 PARTICULATE AND FILLER AS REINFORCEMENTS IN POLYMERS

Particulate fillers and additives are added to polymer matrices to alter the physical and mechanical properties of the polymer. Because of their adaptability and endurance at high temperatures, polymer materials that are reinforced with different fillers and additives are preferred. Commercial and military airplanes and electrical products like heaters and electrodes have all utilized polymers and the polymer-matrix composites supplemented with various sorts of filler particles [37].

Particulate fillers can be described as powdered compounds with a size smaller than 100 microns that are added to polymers to lower costs, enhance processing, and/or change one or more characteristics. Filler addition may affect at least 30 characteristics, thus adding a filler may be done to enhance modulus, but along with that, most of the other properties are modified as well, potentially in a negative way, even if that was not the intention [38].

The need for fillers to be added to polymers are: increase heat resistance while lowering costs, boost the stiffness, reduce creep, reduce shrinkage during molding or polymerization by lowering the exotherm of cure, alter electrical characteristics, reduce the risk of fire, change the specific gravity (density), and change flow. Fillers are also added to increase the compressive strength of the structural element, to improve abrasion resistance by increasing lubricity, reduce or increase permeability, strengthen tensile and flexural properties, increase impact strength, enhance the dimensional stability, improve thermal conductivity, improve processability, increase the moisture resistance, and improve degradability. To modify adhesion to itself or to other substrates, to change the color, opacity, and sheen of the object—these are the situations in which particulate fillers can be used to get the required results.

4.5.1 Particulate filler used by polymer type

Unsaturated polyester resins (UPR)—viscous liquids produced by dissolving low molecular weight polymers in vinyl monomers, primarily styrene—are utilized. This permits the resin to be easily molded or shaped into the required shape before hardening to its final stiff form. UPR is often used in fiberglass-reinforced goods such as car panels, shower stalls, boat hulls, and autobody components [39]. Mineral fillers are frequently used in these in combination with glass fibers.

Epoxy is a somewhat costly polymer that is primarily utilized for its transparency and clarity. The major application of particle fillers in epoxy is in printed circuit boards, as they require fillers that are thermally inert and electrically non-conductive. Fillers include magnesia, alumina, and specialty aluminosilicates. It's also used in flooring and solid surface applications that require hard fillers like quartz [40].

4.5.2 Natural and organic mineral-based particulate fillers

Calcium carbonate fillers are appropriate for a wide range of polymer applications, with global usage exceeding ten million tons per year. Calcium carbonate, or CaCO3, is a common chemical molecule found in nature. Fillers made of calcium carbonate are finely divided forms of this chemical that may be found in a variety of minerals [41]. GCC (ground calcium carbonates) is a common abbreviation for natural calcium carbonates, while PCC stands for synthetic calcium carbonates (precipitated calcium carbonate). GCC is a popular component in thermoplastics, particularly PVC and polyolefins. GCC levels are generally 2–40 phr in rigid PVC (unplasticized or u-PVC); however, low-grade goods with up to 100 phr can be found. Extrusions, such as pipelines, cable management systems, vinyl sidings for doors and windows, and other construction profiles, are the most common uses [41].

Clays of various varieties are readily accessible across the globe and have been utilized in polymer composites, particularly elastomer-based composites. But in comparison to calcium carbonates and talcs, China and ball clays have limited applicability in thermoplastic and thermoset applications but their limitations can be mitigated through calcination and careful treatment. This is due to multiple factors, including poor color retention and heat aging, particularly in PP [39]. Non-black fillers such as China and ball clays are frequently utilized in the rubber industry. When it comes to the influence of fillers on elastomer reinforcement, Skelhorn et. al. described in their study that there are four levels to consider: reinforcing, semi-reinforcing, non-reinforcing, and diluent [39, 42]. The tensile and tear strengths of non-crystallizing elastomers, such as styrene-butadiene rubber (SBR), are utilized to determine reinforcement [43]. At low loadings, reinforcing fillers show a quick peak in performance as well as an increase in wear resistance.

Mica is a platelet-shaped reinforcing filler used in polymers to increase heat deflection temperature and reduce warpage. Types of micas include: muscovite (white to gray), lepidolite (lithium-mica, white to gray), phlogopite (brownish depending on the iron content), biotite (dark brown to black), and other micas. Their use in polyamide is mostly in the automobile sector for under-the-hood applications.

Rice husks, also known as hulls, are the outer layer of rice grains that protect the rice seeds throughout the growth process. They are rich in lignin and amorphous silica, making them tough to digest for humans. Hence, they are discarded and burnt as a fuel source. The product of the ash that is created is called rice hull ash (RHA) and is mainly composed of silica. With almost three million tonnes of silica produced annually around the world, RHA silica is considered a significant waste stream. The ash is pulverized to a particle size of 0.1–2 m, a specific surface area of around 30 m^2/g, and an oil absorption of 30–40 ml/100 g [44].

The particle size and specific surface area of this substance greatly influence its uses. It's too fine to be used as a filler in most thermoplastic and thermoset products. It is more appropriate for usage in elastomers and has a wider range of applications. Maryoto et al. studied the role of rice husks for an alternative source of energy for the cement making process and concluded that rice husks may be used to substitute coal in the cement industry since the calorific content of rice husks is 1 to 2 that of coal [45]. Further along in their study, Deotale et al. revealed that replacing 20% to 30% of cement content with rice husk ash mixed with fly ash increased concrete compressive strength [46].

4.5.3 Synthetic particulate fillers

Carbon black refers to a group of small, usually amorphous or para crystalline carbon particles that have grown together to form aggregates of varied sizes and forms. It's a substance that is formed due to incomplete combustion of heavy petroleum products that include FCC tar, coal tar, ethylene cracking tar, and vegetable matter. Carbon blacks are primarily utilized in tires and other rubber goods as reinforcing fillers. The reinforcing effect is influenced by the interaction between elastomer molecules, carbon black particles, and the elastomer matrix [47]. Carbon black affects elastomer reinforcement by boosting the mechanical strength as well as the wear and abrasion resistance, according to a book by Wolff et al. [48]. The material makes the rubber become stiffer and less easily deformable as a result of the stresses.

Precipitated and fumed silicas are amorphous silicas with very tiny particles and a high specific surface area that are utilized as fillers. Fumed silica is different from silica fume. Silica fume is a by-product of the production of elemental silicon or silicon alloys in electric arc furnaces, and it is used as a cement component (pozzolan). Many silicone elastomer compositions require fumed silicas as a component. Precipitated silicas and silicates, as well as fumed silicas, have a long and successful history as specialized effect fillers for polymer applications, particularly in elastomers and also surface coatings. Precipitated silicas are ideal for elastomer applications because of their small size and high specific surface area. Similar to carbon blacks, they can provide significant amounts of reinforcing, but with a distinct property profile.

4.5.4 Nanofillers

Nanoparticles are often defined as having at least one dimension in the range of 1–100 nm and all other dimensions more than 100 nm. Nanoparticles are described as those with three nano dimensions, nanofibers have two, and nanoplates have one. The stiffest known natural substance is graphite, which has a Young's modulus of about 1,000 MPa. It's a layered mineral made up of graphene layers piled on top of one other. The graphene layers are very strong and rigid, making them perfect for use as nanoplates. Expandable graphite is most commonly used in polymers nowadays as an intumescent fire-retardant ingredient [49]. These graphite compounds are designed to stay chemically stable during the polymer manufacturing process and only exfoliate if they catch fire.

Graphene, on the other hand, has generated a lot of excitement. It's made out of a single graphitic carbon layer, and it's a realistic two-dimensional substance because it's just one atom thick. The strength and stiffness of this material, which are measured at 130 GPa and 1 TPa (terra Pascal), respectively, have piqued people's curiosity and is believed to be the world's strongest substance [49]. Graphene is also the significant primitive unit of carbon nanotubes and nanocarbon fibers.

4.6 ISSUES WITH POLYMERS

Fiber-reinforced polymer (FRP) has seen rapid growth in recent years due to its several advantages and applications. However, there are still many areas where they are lacking, and literature suggests many reasons for this. The main issues are related to cost, structural performance, and durability.

In terms of cost, fiber composite materials are more expensive than conventional materials, whether measured in terms of weight per unit or force-carrying capacity. Factors that contribute to the high material cost include high cost of raw materials and manufacturing, the use of imported materials, and occasional low availability of stock. FRP has a high specific stiffness and high specific strength when it comes to structural performance and durability. In some applications, this helps in the creation of low-self-weight structures. However, low self-weight can cause over-designing of structures for strength, which in turn makes them expensive, and cost saved in other processes becomes insignificant.

According to the research, the chemical consistency of lignin, cellulose, and hemicellulose fibers may have a negative impact on cement hydration, due to their solubility in water [19]. They also exhibit a linear elastic response in tension until failure, or brittle failure. They exhibit poor shear resistance, and fire resistance. When they bend, they lose a drastic amount of strength and are vulnerable to stress-rupture effects.

Although the cost of repair supplies is lower and the volume of materials required is lower, a relatively high level of labor attention is required, which is highly costly. The feasibility of making a sustainable material is still under improvement, since considerations such as energy, material resources, and environmental concerns are hindrances. Since many of the restrictions have been identified, numerous solutions have evolved as a result of study, and future advancements are anticipated. As market share and demand for products increase at a rapid pace, the unit cost of FRP reinforcements is predicted to drop dramatically.

4.7 APPLICATIONS

Fiber-reinforced polymer is now being studied at many research labs and organizations around the world. In general, FRP is more expensive than conventional materials; however, there are few instances where FRP reinforcements are both cost-effective and reasonably applicable. Even when certain structural parts have been extensively damaged due to load specifications, it is possible to improve their strength by using bonded FRP sheets or plates in the repair and strengthening of concrete structures. In addition, FRP meshes or textiles or fabrics are used in thin cement products [50]. Along with enhanced properties, FRP is becoming more sustainable and environmentally friendly. This can reduce the huge amount of waste generated from industries such as manufacturing and construction.

4.7.1 Uses of fiber composite polymer materials in construction

One of the most prevalent applications of polymer fiber composites in the construction sector is the restoration of old buildings and structures [2]. In reinforced and stressed concrete, the material is also utilized to replace steel, and in extremely rare situations, new civil constructions are made nearly completely from fiber composites. Over the last three decades, FRP materials have been used to construct a small number of new load-bearing engineering structures. Examples include pedestrian and vehicle bridges, curved roofs, and decking material in bridges [51]; these materials have also been implemented in building systems; adjustable rooftop cooling towers; energy-absorbing roadside curbs; industrial, chemical, and offshore access platforms [52]; power poles and light poles; electricity transmission towers [2]; and marine structures like sea walls and dock fenders.

Application in fire-resistant concrete: For many years, fiber-reinforced inorganic polymer (FRiP) composites have been extensively used to reinforce concrete structures, and recent research has integrated inorganic/cementitious materials in the production of FRP composites. The fire resistance of the FRP composite structure is improved by replacing the epoxy with phosphate cement-based FRiP [53]. Studies shown by Fang et al. indicated that when exposed to fire, FRiP retains approximately 47% of its strengthening efficacy [54].

Application in concrete slabs: Carbon epoxy and E-glass epoxy composite systems restored the original capacity of damaged concrete slabs, which resulted in an astonishing improvement of more than 540% in the strength of the restored slabs for both unreinforced and RC slabs in a comparative study on damaged concrete slabs. Furthermore, unreinforced specimens demonstrated a 500% increase in structural capacity for retrofitting applications when utilizing FRP systems, whereas steel-reinforced specimens showed a 200% increase.

4.7.2 Applications in other fields

FRP composites and materials are used in manufacturing in the following sectors [35]:

- Aerospace structures: Polymer composites have mostly been pioneered by the military aviation sector. The use of composites in commercial aircraft is progressively growing. Many structural components of the space shuttle and satellite systems are made of graphite/epoxy.
- Marine: Boat hulls and bodies, canoes, kayaks, and other various lightweight category vessels.
- Automotive: Automotive components include body framework, bumpers, driveshafts, doors, and race vehicle bodywork.
- Sporting equipment: Golf equipment and tennis rackets, skis, fishing rods, etc.
- Defense: Bulletproof body armor and vehicle armor.
- Biomedical applications: Medical implants, orthopedic devices, and X-ray tables are examples of biomedical applications.

4.8 CONCLUSION

Polymer composites can be described as materials that exist in different phases, which are manufactured by combining polymer resins like polyester, fillers, epoxy, reinforcing fibers, and vinyl ester, to generate a bulk material that has superior characteristics from the individual basic components. Fillers are frequently used in materials to add bulk, reduce bulk density, lower cost, or provide esthetical characteristics. The addition of fibers to the polymer strengthens it and enhances mechanical characteristics like stiffness and strength. Glass, aramid, and carbon high-strength fibers bear the bulk of the load, while the polymer resin shields and connects the fibers into a unified structurally sound unit. Fiber composite materials are the general name for these types of materials. Since they represent the majority of applications in the most advanced sectors, fiber-reinforced composites have become one of the most auspicious and effective types of composites.

Natural and synthetic fibers are two types of fibers that may be used to make fiber-reinforced composites. Natural fibers are inexpensive and biodegradable, making them ecologically beneficial, whereas synthetic fibers give additional rigidity. Polymer composites have been widely used in the construction industry for a long time in non-structurally critical applications such as bathtubs and vanities, cladding, decorating, and finishing. The construction industry was the world's second-largest user of polymer composites in 1999, accounting for 35% of global demand, according to Weaver et al. In recent years, the construction industry has increasingly considered fiber composite materials for structural load-bearing applications. They've established themselves as a feasible and cost-effective option for the restoration and refurbishment of currently present civil structures, as well as a steel substitute in R.C.C. structures and, to a lesser extent, new civil structures.

The classifications of composite materials, as well as the features of their constituent variants, have been explored to better understand the possibilities of diverse composite materials in various areas. Despite the fact that both natural and synthetic fibers are useful in a variety of applications, new research has demonstrated the superior performance of hybrid fiber-reinforced composite materials, which incorporate the advantages of both natural and synthetic fibers. Individual materials may be substituted with composite materials for specialized purposes in a variety of industries based on the required property improvement. Composite constructions have increased the material's strength and rigidity while significantly reducing its weight. Composites have also exhibited many impressive features, including impact resistance, wear resistance, corrosion resistance, and chemical resistance; however, these properties vary depending on the material's constitution, fiber variant, and manufacturing process.

Composite materials are used in a variety of applications as mentioned in this chapter, depending on the characteristics required. Future studies will focus on developing novel composite structures that incorporate a multitude of variations, as well as new production and manufacturing processes.

REFERENCES

1. F. Campbell, "Introduction to composite materials," in *Structural Composite Materials*, Ed Kubel, Ed., ASM International®, Ohio, 2010, pp. 1–29.
2. M. Humphreys, "The use of polymer composites in construction," in *CIB 2003 International Conference on Smart and Sustainable Built Environment*, Queensland, Brisbane, 2003.
3. A. Ticoalu, T. Aravinthan, and F. Cardona, "A review of current development in natural fiber composites for structural and infrastructure applications," in *Southern Region Engineering Conference*, Toowoomba, 2010.
4. A. A. 544-E, "Mechanical Properties Sub-Committee Meeting," Denver, Colorado, in *Committee, 544, NV*, 2014.
5. K. L. Pickering, G. W. Beckermann, and S. N. Alam, "Optimising industrial hemp fibre for composites," *Composites Part A: Applied Science and Manufacturing*, vol. 38, no. 2, pp. 461–468, 2007.
6. P. K. Mehta and P. J. Monteiro, *Concrete: Microstructure, Properties, and Materials*, Fourth Edition, New York, Chicago, San Francisco, Athens, London, Madrid, Mexico City, Milan, New Delhi, Singapore, Sydney, Toronto: McGraw-Hill Education, 2014.
7. A. Shalwan and B. Yousif, "In State of Art: Mechanical and tribological behaviour of polymeric composites based on natural fibres," *Materials & Design*, vol. 48, pp. 14–24, 2013.
8. Y. Xie, C. A. Hill, and Z. Xiao, "Silane coupling agents used for natural fiber/polymer composites: A review," *Composites Part A: Applied Science and Manufacturing*, vol. 41, no. 7, pp. 806–819, 2010.
9. R. Rowell, "1- Natural fibres: Types and properties," in *Properties and Performance of Natural-Fibre Composites*, K. L. Pickering, Ed., Woodhead Publishing, Madison, Wisconsin, 2008, pp. 3–66.
10. H. Ku, H. Wang, and N. Pattarachaiyakoop, "A review on the tensile properties of natural fiber reinforced polymer composites," *Composites Part B: Engineering*, vol. 42, no. 4, pp. 856–873, 2011.
11. N. M. Nurazzi, M. M. Harussani, and H. A. Aisyah, "Treatments of natural fiber as reinforcement in polymer composites—a short review, " *Functional Composites and Structures*, vol. 3, no. 2, 2021.
12. V. Fiore, G. Bella, and A. Valenza, "The effect of alkaline treatment on mechanical properties of kenaf fibers and their epoxy composites," *Composites Part B: Engineering*, vol. 68, pp. 14–21, 2015.
13. M. Jawaid and H. P. S. A. Khalil, "Cellulosic/synthetic fibre reinforced polymer hybrid composites: A review," *Carbohydrate Polymers*, vol. 86, pp. 1–18, 2011.
14. P. Balakrishnan, M. J. John, and L. A. Pothen, "Natural fibre and polymer matrix composites and their applications in aerospace engineering," 2016.
15. M. N. Norizan, K. Abdan, and M. S. Salit, "Physical, mechanical and thermal properties of sugar palm yarn fibre loading on reinforced unsaturated polyester composites," *Journal of Physical Science*, vol. 28, no. 3, pp. 115–136, 2017.

16. M. Cai, H. Takagi, and A. N. Nakagaito, "Effect of alkali treatment on interfacial bonding in abaca fiber-reinforced composites," *Composites Part A: Applied Science and Manufacturing*, vol. 90, pp. 589–597, 2016.
17. S. H. Kamarudin and L. C. Abdullah, "Thermal and structural analysis of epoxidized jatropha oil and alkaline treated kenaf fiber reinforced Poly(Lactic Acid), Biocomposites," *Polymers*, vol. 12, no. 11, p. 2604, 2020.
18. T. B. Yallew, P. Kumar, and I. Singh, "Sliding wear properties of jute fabric reinforced polypropylene composite," *Procedia Engineering*, vol. 97, pp. 402–411, 2014.
19. F. Kesikidou and M. Stefanidou, "Natural fiber-reinforced mortars," *Journal of Building Engineering*, vol. 25, 2019.
20. K. Bilba, M. Arsene, and A. Ouensanga, "Sugar cane bagasse fibre reinforced cement composites. Part. I. Influence of the botanical components of bagasse on the setting of bagasse/cement composite," *Cement and Concrete Composites*, vol. 25, no. 1, pp. 91–96, 2003.
21. C. RSP. and P. O. T. F. R. I. Symposium, Fibre reinforced cement and concrete, vol. 97, R. Swamy, Ed., CRC Press, London, 1992, pp. 31–47.
22. H. Hajiha and M. Sain, "The use of sugarcane bagasse fibres as reinforcements in composites," in *Biofiber Reinforcements in Composite Materials*, 2015, pp. 525–549.
23. R. Wirawan and S. Sapuan, "Properties of sugarcane bagasse/poly (vinyl chloride) composites after various treatments," *Journal of Composite Materials*, vol. 45, no. 2011, pp. 1667–1674, 2010.
24. K. G. Satyanarayana, J. L. Guimarães, and F. Wypych, "Studies on lignocellulosic fibers of Brazil. Part I: Source, production, morphology, properties and applications," *Manufacturing*, vol. 38, no. 7, pp. 1694–1709, 2007.
25. Z. Zheng and D. Feldman, "Synthetic fibre-reinforced concrete," *Progress in Polymer Science*, vol. 20, no. 2, pp. 185–210, 1995.
26. A. J. Al-Tayyib, M. M. Al-Zahrani, M. M. Al-Zahrani, and G. J. Al-Sulaimani, "Effect of polypropylene fiber reinforcement on the properties of fresh and hardened concrete in the Arabian Gulf environment," *Cement and Concrete Research*, vol. 18, no. 4, pp. 561–570, 1988.
27. R. B. Seymour and R. S. Porter, "*Manmade fibers: Their origin and development*," Elsevier Applied Science, London, New York, 2017.
28. D. C. Hughes, "Stress transfer between fibrillated polyalkene films and cement matrices," *Composites*, vol. 15, no. 2, pp. 153–158, 1984.
29. P. Soroushian and A. B. Khan, "Mechanical properties of concrete materials reinforced with polypropylene or polyethylene fibers," *ACI Materials Journal*, vol. 89, no. 6, pp. 535–540, 1992.
30. J. Beaudoin, *Handbook of Fiber-Reinforced Concrete*. Principles, Properties, Developments and Applications, The National Academies of Sciences, Engineering, and Medicine, New Jersey, 1990.
31. J. K. Patel, N. B. Desai, and J. C. Rana, *Fibre Reinforced Cements and Concretes: Recent Developments*, Elsevier Applied Science, London, 1998.
32. Y. Wang and S. Baker, "An experimental study of synthetic fibre reinforced cementitious composites," *Journal of Materials Science*, vol. 22, no. 12, pp. 4281–4291, 1987.
33. A. Nanni, *Structure Congress*, ASCE, Maryland, 1990.

34. P. L. Walton and A. J. Majumdar, "Cement-based composites with mixtures of different types of fibres," *Composites*, vol. 6, no. 5, pp. 209–216, 1975.
35. D. K. Rajak, D. D. Pagar, P. L. Menezes, and E. Linul, "Fiber-reinforced polymer composites: Manufacturing, properties, and applications," *Polymers*, vol. 11, no. 10, p. 1667, 2019.
36. F. Hanan and M. Jawaid, "Mechanical performance of oil palm/kenaf fiber-reinforced epoxy-based bilayer hybrid composites," *Journal of Natural Fibers*, vol. 17, no. 2, pp. 155–167, 2020.
37. Y. C. Ching, T. U. Gunathilake, and K. Y. Ching, "Effects of high temperature and ultraviolet radiation on polymer composites," in *Durability and Life Prediction in Biocomposites, Fibre-Reinforced Composites and Hybrid Composites*, Woodhead Publishing, 2019, pp. 407–426.
38. C. DeArmitt and R. Rothon, "Particulate fillers, selection, and use in polymer composites," in *Fillers for Polymer Applications. Polymers and Polymeric Composites: A Reference Series*, Springer, Cham, 2017, pp. 3–27.
39. R. Rothon, *Fillers for Polymer Applications*, Springer Science and Business Media, Berlin, 2017.
40. R. Rothon, "Particulate fillers in thermoset plastics 5," in *Fillers for Polymer Applications*, Springer International Publishing AG, Chester, 2017, p. 111.
41. R. Rothon and C. Paynter, "Calcium carbonate fillers," in *Fillers for Polymer Applications*, Springer International Publishing, Cham, 2017, pp. 149–160.
42. Skelhorn, "Chapter 7, Particulate fillers in elastomers," in *Particulate-Filled Polymer Composites*, 2nd Edition, Rapra Technology Limited, Shawbury, 2003, pp. 303–355.
43. R. Rothon, "China Clay or Kaolin, " in *Fillers for Polymer Applications*, Springer International Publishing, Cham, 2017, pp. 161–175.
44. R. Rothon, *Particulate-Filled Polymer Composites*, iSmithers Rapra Publishing, Shrewsbury, 2003.
45. A. Maryoto and G. H. Sudibyo, "Rice husk as an alternative energy for cement production and its effect on the chemical properties of cement," in *MATEC Web Conference*, Jenderal Soedirman University, Indonesia, 2018.
46. R. Deotale, S. Sathawane, and A. Narde, "Effect of partial replacement of cement by fly ash, rice husk ash with using steel fiber in concrete," *International Journal of Scientific & Engineering Research*, vol. 3, no. 6, pp. 1–8, 2012.
47. M. E. Spahr and R. Rothon, "Carbon black as a polymer filler," Roger Rothon, Ed., in *Polymers and Polymeric Composites: A Reference Series*, Springer, Berlin, Heidelberg, 2016, pp. 1–31.
48. S. Wolff and M.-J. Wang, "Carbon black reinforcement of elastomers," in *Carbon Black*, 2nd Edition, Routledge, 2018, pp. 289–355.
49. R. Rothon, "Nanofillers 23," in *Fillers for Polymer Applications*, Springer International Publishing AG, Cham, 2017, p. 436.
50. M. A. Masuelli, "Introduction of fibre-reinforced polymers – polymers and composites: Concepts, properties and processes," in *Fiber Reinforced Polymers-The Technology Applied for Concrete Repair*, InTech, San Luis, 2013, p. 4.
51. FHWA, "https://www.fhwa.dot.gov," 2002. Online. Available: https://www.fhwa.dot.gov/bridge/composite/. Accessed 1 October 2021.
52. J. M. Hale and G. A. Gibson, "Strength reduction of GRP composites exposed to high temperature marine environments," in *Proceedings of ICCM-11*, Gold Coast, 1997.

53. C. Menna, D. Asprone, C. Ferone, F. Colangelo, and A. Balsamo, "Use of geo-polymers for composite external reinforcement of RC members," *Composites Part B: Engineering*, vol. 45, no. 1, pp. 1667–1676, 2013.
54. Y. Fang, P. Cui, Z. Ding, and J.-X. Zhu, "Properties of a magnesium phosphate cement-based fire-retardant coating containing glass fiber or glass fiber pow-der," *Construction and Building Materials*, vol. 162, pp. 553–560, 2018.

Chapter 5

Mechanical testing and its performance

M. Aruna
Al Ghurair University, Dubai, UAE

CONTENTS

5.1 INTRODUCTION

Studies on natural fiber reinforcement composites have focused on both fibers and matrixes originating from renewable resources, and findings indicate further compatibility with environmental conservation issues. Composite materials centered on renewable raw materials from agriculture and biomass are widely utilized, as these products use fewer fossil fuels and thus diminish greenhouse gas emissions associated with petrol-based materials.

Interest in natural fiber-reinforced composite materials is growing, both in terms of industrial purposes and essential research. Researchers, engineers, and scientists now regard natural fibers as an alternative to fiber-reinforced polymer (FRP) composites. Because of natural fibers' ease of use, low cost, good mechanical properties, elevated specific strength, and

DOI: 10.1201/9781003252108-5

83

non-abrasive, eco-friendly, bio-degradability characteristics, they are utilized as an alternate option for conventional fiber such as aramid and carbon.

Polymers synthesized from natural resources have gained considerable research interest in recent years. Environmental and economic concerns have stimulated research in the development of new materials for construction, furniture, packaging, and automotive industries. Particularly attractive are the new materials derived from natural renewable resources that would prevent further stress on the environment such as that caused by the depletion of already dwindling wood resources from forests.

Studies show that mixing materials enhances their properties. Fibers act as a reinforcement to matrices—binding the fibers together, shielding them from the environment and destruction due to handling, and spreading the load. Although matrices alone normally have low mechanical properties compared to those of fibers, the matrix impacts many mechanical properties of the composite. These properties comprise transverse modulus and strength, shear modulus and strength, compressive strength, interlaminar shear strength, thermal expansion coefficient, thermal resistance, and fatigue strength.

Four new models for Young's modulus of concentrated particulate composites are created using a differential scheme, along with the solution of an infinitely dilute dispersion of particles in a solid matrix [2]. While it is true that the high strength of composites is mostly due to fiber reinforcement, the importance of matrix material cannot be underrated as it delivers support for the fibers and assists them in carrying loads. It also gives stability to the composite material. Dispersions appear alongside those of solid composite materials [1].

Because of their light weight and extraordinary mechanical properties, composites are employed in a variety of industries. Composite materials are comprised of two main types of elements such as matrix and reinforcement. The matrix material envelops and supports the reinforcement materials by preserving their relative positions. The resin matrix system acts as a binding in which the fibers are embedded. When too much resin is used, the part is categorized as resin rich. But if there is too little resin, the part is termed resin starved. A resin rich part is more susceptible to cracking due to lack of fiber support; however, a resin starved part is softer because of void areas. Technically, the prime composites are those in which the dispersed phase is in the form of a fiber.

Design targets of fiber-reinforced composites often comprise superior strength and/or stiffness on a weight basis. These characteristics are noted in terms of specific strength and specific modulus parameters, which resemble, individually, the ratios of tensile strength to specific gravity and modulus of elasticity to specific gravity. Fiber-reinforced composites with extraordinarily high-level specific strengths and moduli have been constructed that employ low-density fiber and matrix materials. As noted in

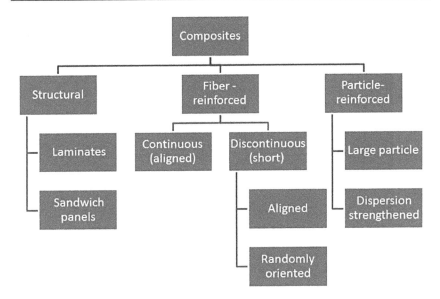

Figure 5.1 Classification of composite types.

Figure 5.1, fiber-reinforced composites are sub classified by fiber length. For short fiber, the fibers are extremely short to produce a considerable progress in strength.

Numerous chemical variations are used to increase the interfacial matrix-fiber bonding resulting in the enhancement of mechanical properties of the composites. Calotropis procera (Figure 5.2) is a species of Calotropis, native to Cambodia, Indonesia, Malaysia, Philippines, Thailand, Sri Lanka, India, China, Pakistan, Nepal, and tropical Africa. Flowers are replaced by a kidney-shaped fruit, 2.7 to 4 inches long. It is green and becomes brown when mature. The fruit contains many seeds embedded in a rough fiber. The fruit is a follicle and when dry, the seeds are dispersed by wind. Fruits are green fleshy follicles; seeds are attached with abundant white coma. The fruit is a grey-green bladdery pod, 8–12 cm long, rounded at the base but shortly pointed at the tip and containing numerous seeds. The seeds are brown flattened, with a tuft of long white hair at one end. The fruit of the plant is green with an ovoid shape.

The flesh of the fruit contains a toxic milky sap that is extremely bitter and turns into a gluey coating resistant to soap. The fruit is large, producing a milky exudate when cut. Seeds are winged, very numerous, each about $8^{-10} \times 5mm$, and attached by broad shiny, white funicles. The testa is clothed in short, erect, and white hairs. Cotyledons are ovate, shaped like a tombstone, much wider than the radical. The fruit of Calotropis procera is oval and curved at the ends of the pods. The fruit is thick, and when opened, is the source of thick fibers used to make rope, and also serves many other uses.

Figure 5.2 Calotropis procera plant.

The plant yields a durable fiber commercially known as bowstrings. The stem is useful for making ropes, carpets, fishing nets, and sewing thread. Fiber from the inner bark was once used in the manufacture of cloth for the nobility. Floss, obtained from the fruit, is used for stuffing purposes. The stem of the milkweed plant has also been used to extract oil and natural rubber. The potential of using milkweed plants as a source of pulp for paper has also been studied. Currently, milkweed plants are being commercially processed for floss used in comforters. Other parts of the plants are also being utilized; however, there are no reports available on the use of the milkweed stems as sources for high-quality natural cellulose fibers.

In a previous work the Calotropis Gigantea (CG) fiber was identified and its physical, chemical, and mechanical properties of CG fibers had been studied. The obtained values of various properties were comparable to the those of other lignocellulosic fibers available in the literature, confirming their potential as reinforcements in polymer matrix composites [3–4].

CG fiber-based polymer composites can thus be developed for numerous applications and can be used as reinforcement in various polymer matrices for developing composite materials. The slight reinforcing effect could be explained with the high stiffness of the cellulose-based particles, which effects a load transfer from the polymeric matrix to the particles. For the maximum strength and elongation, the weakest point within the composite is relevant: the higher the filler content, the more probable the occurrence of a critical agglomeration of particles that can act as a starting point of cracks weakening the composite.

The mechanical behavior of a fiber-reinforced composite basically depends on the fiber strength and modulus, matrix strength, the chemical stability, and the interface bonding between the fiber/matrix to enable stress transfer [6]. Polyester matrix-based composites have been widely used in marine applications, where water absorption is a key factor in degradation of polymer composites. Initiation, propagation, branching, and termination of degradation mechanisms are used to determine the degradation of materials [7].

Composite-reinforcing fibers can be categorized by chemical composition, structural morphology, and commercial function. Natural fibers such as kenaf, ramie, jute, flax, sisal, sun hemp and coir are derived from plants used almost exclusively in polymer matrix composites (PMCs). Aramid fibers [6] are crystalline polymer fibers mostly used to reinforce PMCs. Hence, the mechanical properties of PMCs are located on the composition percentage of composites, which is the most critical aspect of verifying the designed values [9]. The primary function of a reinforcing fiber is to increase the strength and stiffness of a matrix material. The fiber-reinforced composite's significant advantages are high stiffness, light weight, easily recycled material, availability, low manufacturing cost, positive environment effect, and lifetime rupture behavior.

Various types of natural fibers can combine with other mineral fiber to construct composite material. These fibers can be classified as vegetable, animal, and man-made. The main disadvantages of natural fibers are their high level of moisture absorption, poor and interfacial adhesion, and relatively low heat resistance [7–8]. This research indicated significant improvements in the penetration resistance, which came from the improvement of target geometry structure. Major efforts are being performed in the field of polymer science and technology to arrange macromolecular materials based on renewable resources [10].

Researchers conducted an experimental study about the mechanical behavior of epoxy composites reinforced with sisal and glass fiber [11]. Compared to glass fiber composites, hybrid composites have lower tensile strength due to the presence of sisal fibers. This study concluded that the poor interfacial bonding is responsible for low mechanical properties. Experimental investigation on the effect of sisal fiber in cement composites concluded that they are more reliable materials to be used in practice to produce construction materials [5]. In this context, the study of the mechanical properties of Calotropis procera fiber-reinforced epoxy composites is detailed.

Its fiber was extracted physically from the plant's stem and treated chemically, as shown in Figure 5.3. The mercerization was carried out as per

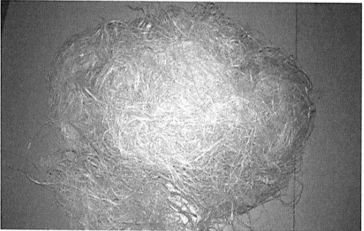

Figure 5.3 Calotropis procera fiber before and after mercerization.

ASTM 1965. Calotropis fibers were immersed in 5 wt% NaOH solutions for 24 to 48 hours, and the results of the alkali treatments on the mechanical characteristics and interfacial adhesion of the fibers in the Calotropis/epoxy composite system were systematically evaluated. The fibers were then washed with distilled water and left to dry at room temperature before being put in an oven for 15 h at 70°C. Any residual NaOH on the fiber surface was deactivated with 2% sulfuric acid throughout 10 minutes. The fibers were eroded with distilled water until gaining a pH = 7. After chemical treatment, the fibers were dried at 60°C for 6 hours. Appropriate drying is important is for superior strength. The samples were made by altering the fiber percentages and epoxy resin.

Hybrid composites are fabricated using raw Calotropis procera stem fiber/glass with varying fiber weight percent 5:0 to 5:3 weight by using the hand lay-up method. Hand lay-up is the easiest and eldest open molding method of the composite fabrication processes. It is a low volume, labor-intensive method fit specifically for large parts, such as boat hulls. Glass or other reinforcing mat or woven fabric or roving is arranged manually in the open mold, and resin is poured, brushed, or sprayed over and into the glass plies. Rolling is done to eliminate entangled air from laminates. Room temperature curing epoxies and polyester are the highly utilized matrix resins. Curing is instigated by a catalyst in the resin system, which toughens the fiber-reinforced resin composite with no external heat. For a high-excellence part surface, a pigmented gel coat is initially put on to the mold surface.

Composite samples were prepared using the molding pattern prepared in mild steel. Wax is applied on all surfaces of the mold, which helps to remove the pattern, and to take out the fabricated composite material easily. As an initial step, catalyzed resin applied on the inner edges of the molding pattern, helps the pattern remain fixed. The measured fiber is spread on the molding box. Epoxy resin is poured on the fiber, and it is dipped within the fiber inside the molding box. Then 5 kN load is applied in a compression molding machine, and the whole setup was left for 24 hours. The obtained composite plate is shown in Figure 5.4. The manufactured specimens were cut as rectangular pieces corresponding to the ASTM standards for performing all tests.

Ever-expanding knowledge regarding natural fibers will enable standardization of the variety of fibers in the market and give designers more responsibility concerning the mechanical and chemical properties.

Figure 5.4 Calotropis procera fiber composite slab.

5.2 MECHANICAL PROPERTIES

5.2.1 Tensile testing

The primary tensile strength test will deliver an indication of the necessary mechanical properties of the composite material. These outcomes can then be utilized as a rapid test to verify the material integrity and guarantee that manufacture conforms to a "fit for purpose" standard. The tensile strength can also be treated to deliver a comparable indication of quality. Factors that impact the tensile response and that should therefore be confirmed include the following: material, methods of material preparation and lay-up, specimen stacking sequence, specimen preparation, specimen conditioning, environment of testing, specimen alignment and gripping, speed of testing, time at temperature, void content, and volume percent reinforcement. Properties obtained from this test method include the following:

- Ultimate tensile strength
- Ultimate tensile strain
- Tensile chord modulus of elasticity
- Poisson's ratio
- Transition strain

The tensile test was accomplished on all the samples using the universal strength testing machine Instron 4485 as per ASTM D3039-76 test standards. ASTM D3039 is a commonly used testing standard for verifying the tensile properties of composite materials. The tension test is normally achieved on flat specimens. The precise value of strains was evaluated with a mechanical extensometer with strain gauge bridge. Two samples from each considered stacking sequence were subjected to laboratory tests. A uniaxial load was applied across the ends. The ASTM standard test advises that the specimens with fibers parallel to the loading direction must be 11.5 mm wide, with the length of the test section as 120 mm. Among these composites there is a substantial increase of tensile strength peak load 276.49 kg/cm^2 for the hybrid sample as the proportion of glass fiber increases to 20 gm. The testing was performed on the samples of Calotropis/glass combinations. Three specimens were tested for each mixture and the median was considered for analysis. The assessed (untreated and treated short fiber) composites as a function of fiber content in tensile, impact, flexural, hardness, and compression properties of fiber-epoxy composites are furnished for the six samples in Table 5.1. Figure 5.5 depicts the tensile tested composite slab.

Open hole tensile testing determines the force necessary to break a composite specimen with a centrally positioned hole. ASTM D5766 is a suitable test procedure for simulating flaws in a composite material component and for producing data where the end-use of the material needs a fastener hole. The filled hole tensile and compression test approaches are illustrated by the

Table 5.1 Test results of the composite specimens

Specimen number	S1	S2	S3	S4	S5	S6
Glass fiber	10 wt%	15 wt%	20 wt%	10 wt%	15 wt%	20 wt%
Treatment		Mercerized			Non-Mercerized	
Tensile strength [kg/cm²]	260.78	275.49	276.49	275	273.4	272.12
Impact strength [Joules]	4.521	6.1	6.559	4.7	4.563	4.67
Hardness [HV]	93.5	95.5	96.9	87.2	93.1	92.3
Flexural Strength [MPa]	108	117	126	31	28	22
Compression strength [kN]	320	440	500	180	200	240

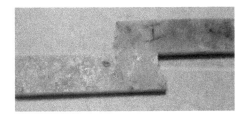

Figure 5.5 Tensile tested composite slab.

ASTM D6742 specification. This technique is similar to open hole testing, though only the material is tested while a fastener is in the hole. Throughout filled hole testing, the material usually withstands a greater force. The clearance of the hole, fastener type, material hardness, and torque may impact the material's strength properties.

5.2.2 Impact testing

An analog Izod/Charpy Impact Tester was utilized for testing the impact properties of the Calotropis fiber-reinforced composite specimen. The equipment had a minimum resolution on each scale of 0.02 J, 0.05 J, 0.1 J, and 0.2 J, respectively. Four scales and corresponding hammers (R1, R2, R3, R4) were provided for all the above working ranges. The Impact energy of 20g weight percentage of Calotropis procera stem fiber-reinforced composites (Sample 3) is 6.559 J, which is better than other samples due to an increase in glass fiber weight % and mercerization. Figure 5.6 illustrates the Impact tested composite slab.

5.2.3 Hardness testing

Hardness is defined as the ability of a material to resist plastic deformation. Generally, the hardness of a composite increases with mercerized fibers. In common, the fiber expands the modulus of combinations, which in turn

Figure 5.6 Impact tested composite slab.

increases the hardness of the fiber. Hardness is a combination of relative fiber volume and modulus.

After 5 wt% NaOH treatment, Calotropis fiber exhibited improved crystallinity, tensile strength, and Young's modulus associated to untreated fibers, and similarly, better interfacial shear strength with an epoxy. Sturdier alkali handlings destructively wedged fiber stiffness and appropriateness for composite applications. Results indicate that moderate alkali treatments (5 wt% NaOH for 2 h) are highly favorable for manufacturing Calotropis fiber-reinforced polymer composites. The test samples with dimensions of 20 × 20 × 3 mm are used for the hardness test conducted on the Vickers hardness testing machine. The purported composite offers superior hardness as 96.9 HV for sample 3 with an increase in the glass fiber by 20 weight %.

5.2.4 Compressive strength testing

During the testing method, compressive force is initiated into the specimen by combined end- and shear-loading. A universal testing machine can be used to test the straight sided composite specimen of a rectangular cross section. Table 5.2 depicts the composite compression test methods. ASTM standards can be used to record the following quantities for the laminate:

- Compressive strength
- Compressive strain
- Compressive (linear or chord) modulus of elasticity
- Poisson's ratio in compression

Compression properties of composites are obstructed by the resin and fiber resin interface and regularly decide the overall design, size, and weight of a part. Composite compression testing is essential for screening new materials and preserving quality control. ASTM D6484 lists employ open hole compression testing to evaluate the compressive properties of a composite specimen with a hole cut in it. The assessment method calls for the use of a side support fixture during testing to avoid Euler buckling, as the only acceptable failure mode is at the hole.

Table 5.2 Composite compression test methods

Test	Specimen	Unsupported gauge length	Findings
ASTM D5467, D7249	24″ × 1″ bonded to a sandwich beam	0″	Fabrication is expensive
ASTM D3410	5.5″ × 1″	0.5″	Large, heavy fixture
ASTM D695	3.13″ × 0.5″	0.5″	Suitable for woven materials and rigid plastics
ASTM D6641	5.5″ × 0.5″ or 1″	0.5″	Most common method, has unsupported length, and is easier to use
ASTM D5467M	24″ × 1″	0″	For unidirectional polymer matrix composite materials using a sandwich beam

The goal of composite compression tests is to drive compressive forces into a uniform gauge area without triggering premature failure or inappropriate breakdown modes in other parts of the specimen. As composite materials evolve, the creation of a universal standard for performing compression testing has presented challenges. The present composite compression test methods are categorized into three classes, differing on how materials are loaded on the test frame: bending, shear, and end-loaded.

(a) Bend compression testing

ASTM D5467 and ASTM D7249 are two of the most common bending test methods for compressive properties of composites. Bend loading employs a sandwich beam assembly for testing, where one face is in compression and the opposite face is in tension. The compressive face is normally built with a thinner face sheet so that the beam breaks first in compression. Once the specimen is constructed, spans, core material, thickness, and adhesives are chosen to confirm that the exact failure mode (compression facing) is achieved. Strain gauges on the compressive face are employed to assess modulus, Poisson's ratio, and strain to failure.

(b) Shear loaded compression testing

Certain techniques employ a shear load to the gauge area of the specimen. One of the first methods, often called Celanese compression, loads the sample through wedges that are clutched by a cone-shaped fixture. The samples are rectangular and normally tabbed. Celanese compression was initially referred to in the ASTM D3410 standard but has since been eliminated. The Celanese samples were commonly limited to 0.25″ in width by the ASTM Standard and 10 mm by ISO 14126. The overall length is 5.5″ with a 0.5″ unsupported length. Even though Celanese compression is mostly considered antiquated, it is occasionally utilized to compare to historical data. For this process, material thickness is a crucial concern and must be within the narrow range specified, which permits proper seating of the wedges with the cones.

The recent form of ASTM D3410 is generally called the IITRI method, which was created to relieve some of the weaknesses in the Celanese method. The IITRI fixture is huge and thick, which creates challenges when used at extreme temperatures. Samplings used for this method may be up to 1″ wide and 5.5″ long with 0.5″ of unsupported length. The wedges load through a rectangular seat, so the complete thickness of the specimen is not the same as with the Celanese method. Still, the material within the gauge area needs to be thick enough to prevent Euler buckling.

(c) End-loading compression testing

Plastics have conventionally been tested employing the end loading methods depicted in ASTM D695. This method allows for the end-loading of a prism that is square or round in cross-section. A dog-bone shaped flat sample may also be utilized and is loaded in a side support fixture that excludes buckling. The Suppliers of Advanced Composites Materials Association (SACMA) agreed and enacted an industry practice of using tabbed specimens with a 0.188″ gauge length. The sections commonly end in a higher apparent compressive strength because of the shorter gauge length. Since all these specimens are end-loaded, the ends must be flat and parallel for significant results. If the samples fail or broom at the ends, the outcomes should be judged invalid. Furthermore, if the level of torque applied to the bolts that ensure the specimen to the fixture is superior, test findings may be invalid. Due to the short gauge length and the inconsistency of friction from the side support fixture, additional consistency of this method will be deliberate.

(d) Combination of shear and end-loading compression testing

Continued fixture improvement resulted in a combined loading compression fixture. ASTM D6641 defines the correct fixture and test method. Identical to the previous methods, the minimal specimen is 5.5″ long with a 0.5″ of unsupported length. By fastening the ends of the fixture between square ends, the sample is loaded with a mixture of end and shear loading. The combined loading compression fixture is reduced and easier to use than the IITRI fixture. Delicately torquing the fixture bolts to an even, specified level is important. Samples are strain gauged with back-to-back gauges to check for Euler buckling. The percent bending is registered with the test data and may be utilized to troubleshoot atypical data. For composite design, values for compression with a hole to reproduce a stress riser is regularly required. ASTM D6484 is the correct process for finding those values.

From the various sections of the prepared composite, the specified dimension as $200 \times 40 \times 12.5$ mm were considered for the compression test, which was performed on a computerized universal testing machine with a power of 20 tons with ASTM standard. According to the literature, six specimens are tested, and the mean value is tabulated, for which all six specimens are examined with a crosshead speed of 2 mm/min at room temperature (25°C).

5.2.5 Flexural strength testing

Flexural strength was basically impacted by the fibers' application. Increased quantity of used fibers led to better flexural strength but not exactly according to their total dosage. The flexural strength of the composite plate was scrutinized by experimental results. Test specimens with the dimensions of 200 mm × 45 mm × 12 mm were used, and the flexural test was finished using a three-point bending set-up according to the ASTM D790 standard. The span length of the sample thickness ratio was retained at 16:1. Six specimens were analyzed for each treating state and the average reported.

a) Thermal properties of composites

Certain raw materials used in the manufacture of composite materials can be undesirably affected by temperature and humidity. It is therefore necessary to make sure they are stored within highly controlled environmental conditions. When the raw material is delivered to the product or parts manufacturer, mandatory checks should be conducted to make sure that the batch properties of each material have not been adversely affected in transit. This inspection process can be particularly significant in warmer, more humid climates or during a long courier transportation trip. Materials delivered from one country to another can potentially be exposed to temperatures that could cause the materials to absorb moisture.

b) Differential Scanning Calorimetry (DSC)

Differential Scanning Calorimetry is a test that confirms the material's thermal properties. DSC is one of the most significant test types when examining quality control. It is a thermo-analytical method in which the difference in the amount of heat involved to raise the temperature of a sample and reference is measured as a function of temperature. Together the sample and reference are held at nearly the same temperature all through the experiment.

Comparison of previous results with tested outcomes is possible where the seller has built up an adequately huge database. It presents data on the physical structure of the material via vital thermal evolutions such as glass transition temperature (Tg), melting temperature (Tm), crystallization temperatures, percent crystallinity, enthalpy of melting and crystallization, specific heat capacity (Cp), and oxidation induction time (OIT). This process determines several characteristic properties of a sample and can examine fusion and crystallization events as well as glass transition temperatures. Additionally, it can also be employed to study oxidation, as well as other chemical reactions.

c) Oxidation Induction Time

Oxidation Induction Time is an assessment of the time for which an antioxidant system present in the polymer inhibits oxidation. OIT measures the level of thermal stabilization of the material tested, while the material is kept at a particular temperature or heated at a constant rate in an oxygen atmosphere. Antioxidants are frequently added during the formulation to improve the aging triggered by oxygen and to enhance the material's lifespan. OIT can also be used to investigate the degree of polymer aging throughout the material's lifetime due to exposure to heat, oxygen, light, and radiation.

d) Dynamic Mechanical Analysis (DMA)

Dynamic Mechanical Analysis is a technique that provides information on the material's physical structure via its viscoelastic mechanical properties. The test captures the material's response to a sinusoidal force during a temperature or frequency sweep. DMA can be used to determine the mechanical properties (mechanical modulus or stiffness and damping) of the composite and important thermal transitions of the adhesive, such as the glass transition temperature and the degree of cure of polymer and composite materials. DMA and DSC tests can also be combined to provide an easy, quick test program to compare proposed alternatives to new raw material.

e) Thermogravimetric Analysis (TGA)

Thermogravimetric Analysis is an analytical method that ascertains a material's thermal stability or assesses information on the material's chemical and physical structure via thermal decompositions. TGA compiles a report on the temperature and rate of decomposition of materials and the quantity of volatiles and fillers they comprise. With advanced analysis software, characteristic temperatures such as melting points and decomposition temperatures can also be evaluated. The weight modifications of polymeric materials can be initiated by decomposition and oxidation reactions as well as physical processes such as sublimation, vaporization, and desorption.

f) Thermomechanical Analysis (TMA)

Thermomechanical Analysis is employed to view the material's coefficient of thermal expansion (CTE and also widely used to identify the thermal properties of polymeric materials. Using negligible force at a range of temperatures, TMA can find a variety of thermal and mechanical properties, including thermal expansion. This technique facilitates the distortion of a material under a constant load to be assessed while the material is exposed to a controlled temperature program. This is crucial to quantify because the expansion or contraction of a material under different circumstances can produce problems during the manufacture of new parts. Quantification at an early stage can help to resolve any possible future-fit and assembly issues.

g) Physical properties of composites

Density, volatile material percentage, resin percentage, fiber percentage, and percentage of porosity or void content are significant basic material physical properties of composites. They must be evaluated through fundamental content testing before the raw material reaches the latter phase of production. Exceptionally strict limits regarding expected results are normally set by the end-user. Quantification of the material density is important, as it provides information on the level of crystallinity available in it. Molecules in the crystalline phases compress together more tightly than those in the amorphous phases, giving the material greater density. Fiber areal weight and prepreg areal weight must be tested when signing off on the quality of newly purchased raw material. This is to make sure that the mix ratio between the fiber and resin in the prepreg is optimized to provide ideal end properties.

h) Fourier transform infrared spectroscopy (FTIR)

Fourier transform infrared spectroscopy supports evidence on the chemical structure of a material. This technique is utilized to acquire an infrared spectrum of a solid, liquid, or gas, which can be used to characterize the component. The distinct chemical bonds between atoms provide several signals, with each material having a distinctive FTIR spectrum. FTIR can be employed as a material quality control test to make sure the chemical composition is correct. It can also be applied to explore an unspecified material by linking the unknown sample spectrum with stored library spectra. Any expected heat damage can be identified using this technique, as the chemistry of the material may change after being subjected to elevated heat.

i) Gel Permeation Chromatography (GPC)

Gel Permeation Chromatography, also known as Size Exclusion Chromatography (SEC), is a technique that splits up polymer molecules in the sample corresponding to chain length. The polymeric molecular weight distribution (MWD) can be determined using GPC. It is crucial to compute this individual material property because it correlates to the physical characteristics of the finished material, such as tensile strength and crack propagation qualities. The polymer producer will need an optimized weight range where the ideal properties are neutral. Illustrations of such properties consist of viscosity, which is essential for processability, sample strength, toughness, and crack resistance.

j) Composite drop weight impact testing

Drop weight impact testing evaluates the damage resistance of the material against various impact energies. The process entails dropping a known mass from a known height, perpendicular to the face of a composite specimen. Impact testing offers knowledge about the resistance of a composite to impact damage and can be used in conjunction with Compression After Impact (CAI) testing to decide damage tolerance. CAI results show how an impact affects the subsequent compression strength of the material. Non-destructive inspection practices following impact assist in measuring the depth and size of the damaged area. The Barely Visible Impact Damage (BVID) index serves as a measure of the force a composite can withstand.

5.3 FATIGUE AND CREEP

As composites become more widely used across industries, it's important to understand their durability and damage tolerance. Composite fatigue testing evaluates stiffness and strength reductions in composite materials and the properties of the material when exposed to cyclic loading over time. The fatigue behavior of one constituent may be significantly affected by the presence of other constituents and the interfacial regions between the fibers and matrix. Composite fatigue testing differs from fatigue testing of metals because composite fibers are oriented, which means the fatigue properties hinge on direction, layup, and failure mode.

Normally, fatigue properties are not a serious concern for fiber-dominated materials, as fatigue cracks dissipate upon reaching the next fiber. This technique can be applied in the analysis of fatigue impairment in a polymer matrix composite such as the existence of microscopic cracks, fiber fractures, or delamination. The sample's residual strength or stiffness, or both, may alter owing to these damage mechanisms. The loss in stiffness may be quantified by discontinuing cyclic loading at selected cycle periods to achieve the quasi-static axial stress-strain curve employing modulus determination procedures found in ASTM test method D3039/D3039M [12]. The circular and rectangular notch designs indicate that the notch depth has no influence on the tensile strength values of a pinned laminate [13]. The loss in strength correlated with fatigue damage may be resolved by discontinuing cyclic loading to acquire the static strength using test method D3039/D3039M. This process is restricted to unnotched test samples exposed to constant amplitude uniaxial in-plane loading where the loading is described in terms of test control parameters. Duplicate tests may be applied to acquire a distribution of fatigue life for special material types, laminate stacking sequences, environments, and loading conditions. Assistance in statistical analysis of fatigue life data such as principle of linearized stress life (S-N) or strain-life (ε-N) curves are practiced as per ASTM standards. Different methods of composite fatigue testing for evaluating the fatigue properties of composite materials are

- tension-tension
- tension-compression
- compression-compression

- bending fatigue
- fatigue crack growth

Based on this method, fatigue tests may be achieved in-plane in the fiber governed direction, or out of plane with interlaminar loading to assess damage and damage growth. Composite fatigue testing labs can have multiple load frames equipped with aligned hydraulic grips for a range of load capacities to support the different methods.

Regardless of their high strength and durability properties, composite materials are still vulnerable to in-service damage. It is significant to realize composite damage tolerance to mitigate risk and confirm the material is fit for use. Composite structures may be broken during manufacture or in-service, and those damages aren't always detectable during subsequent inspections. There is a variety of test methods that provides manufacturers with more insight into their composite materials' damage tolerance and resistance and how that damage may spread during service. Two conventional techniques are drop weight impact testing and fracture toughness testing.

5.4 WEAR AND EROSION STUDY

Wear is defined as the progressive loss of materials from the contacting surfaces in relative motion. The wear test (dry sliding wear test) was performed on the fabricated composite, which is a type of adhesive wear. Wear is triggered between the two elements, which is sliding under the practical load and environment. Wear test was done using the Pin-on-disc apparatus. The disc was made of stainless steel with 50 mm diameter and 10 mm thickness. The hardness of the drive was 70 HRC. The test was performed for a particular test duration by utilizing load and sliding velocity. The face of the specimen was kept perpendicular to the contact surface. Prior to testing, the surface of both the sample and the disc were polished with a soft paper soaked in acetone. The initial and final weight of the sample was measured using an electronic digital balance. The shift in the initial and final weight is the measure of weight loss.

The erosion and corrosion percentage of the samples should be estimated to determine the tribological performance of the polymers. Extensive study has been made on the erosive response of polymers with the addition of filler material and fiber. The tests were conducted using a jet at 30 m/s and impinging angle 90°. Exposure time was the major factor in determining erosion valuation, which was from three to six and six to 12 hours at room temperature. The samples' roughness data were recorded before and after the tests by image processing software, and damping properties were determined by mechanical spectroscopy using an inverted torsion pendulum as a function of temperature (150–400 K), frequency (0.01–3 Hz), and strain amplitude [14].

All samples were cleaned with acetone and weighed with an electronic balance with a sensitivity of 0.01 mg prior to the erosive wear tests. The room temperature erosion test was used with irregular silica sand particles of 150 ± 15 μm and powered by a static pressure of 1–3.5 bar. Composite samples were mounted in the specimen holder and subjected to a particle flow at given impingement angles between 30° and 90° to the specimen surface, and erosion was measured after one, two, and three minutes. A scanning electron microscope (SEM) then measured the morphology of the eroded surfaces. The erosion of the fiber is mainly produced by micro-cracking or plastic deformation owing to the impact of silica sand, which increases as kinetic energy loss rises. Kinetic energy loss is maximum at an impingement angle of 90°, where erosion rates are maximum for brittle materials.

Typically, thermoplastic matrix composites show ductile erosive wear (plastic deformation, ploughing, and ductile tearing), while thermosetting matrix composites erode in a brittle manner (generation and propagation of surface lateral cracks). Nevertheless, this failure categorization is not absolute because the erosion behavior of composites depends on the experimental conditions and composition of the target material.

The impingement angle is one of the most important constants in erosion behavior. When the erosive particles hit the target at low angles, the impact splits it into two components: one parallel (Fp) to the surface of the material and the other vertical (Fv). Fp controls the abrasive, and Fv is responsible for the impact phenomenon. As the impact angle shifts towards 90°, the effects of Fv become marginal. With regular erosion, all existing energy is dissipated by impact and micro cracking, while at oblique angles—due to the pivotal role of the Fv—the damage is caused by micro-cutting and micro-ploughing. Scanning probe microscope image process software offers a reliable means to estimate erosion volume loss.

5.5 FAILURE ANALYSIS

Failure analysis is vital to product development and system improvement. It not only helps us learn from previous failures, but also how to avoid them in the future. Failure analysis is a multifaceted, holistic approach that determines how and why a material or product failed. The initial stage of any investigation involves an in-depth discovery phase of the circumstances surrounding the failure and any relevant background information, including environmental factors, type of application, service life, and pertinent design information.

Many methods and tools are available to inspect the failed part. Pinpointing the root cause and linked responsibility of any failure are the primary goals, but the added value is in preventing future happenings. Many criteria for failure have been used over the past few decades; the most common failure criterion for composite materials is the Tensor Polynomial

Criterion proposed by Tsai and Wu. Further popular and known failure criteria involve those proposed by Tsai-Hill, Azzi-Tsai, Hoffman, and Chamis. These conditions don't consider the heterogeneous nature of a lamina, and an increase in mechanical and thermal loads may result in laminate failure. The laminate failure, however, may not be catastrophic. It is likely that one layer of the composite initially fails, which then leads to the failure of all plies. Failed plies may impact depending on the stiffness and strength of the laminate.

5.5.1 Progressive failure by maximum stress criterion

Composite sandwich panels are made of two thin, stiff face sheets bonded to a moderately thick, lightweight core. To determine the durability and performance of a sandwich panel, the entire construct must be tested together. The initiation and propagation of intralaminar failure mechanisms—matrix tension, matrix compression, fiber tension, and fiber compression—are detected, and the stiffness degradation at a particular integration point in the material is studied. Modeling the progressive failure of composite laminas entails specific failure algorithms that are not only computationally effectual but are also able to forecast load-deformation characteristics and to eventually create the suitable failure load. Failure location at the middle and Maximum Stress Criterion is considered. Degradation factors are considered as G12: 0.10, NU12: 1.00, E22: 0.01, E11: 1.00. The following matrices are generated and then inverse of each matrix is considered.

$$
\text{Matrix A} \quad
\begin{matrix}
2.27E+09 & 4.53E+07 & 0.00E+00 \\
4.53E+07 & 1.92E+08 & 0.00E+00 \\
0.00E+00 & 0.00E+00 & 3.03E+10
\end{matrix}
$$

$$
\text{Matrix B} \quad
\begin{matrix}
0.00E+00 & 0.00E+00 & 0.00E+00 \\
0.00E+00 & 0.00E+00 & 0.00E+00 \\
0.00E+00 & 0.00E+00 & 0.00E+00
\end{matrix}
$$

$$
\text{Matrix D} \quad
\begin{matrix}
2.15E+10 & 4.37E+08 & 0.00E+00 \\
4.37E+08 & 1.87E+09 & 0.00E+00 \\
0.00E+00 & 0.00E+00 & 4.53E+10
\end{matrix}
$$

The Max Stress Criterion recognizes three possible modes of failure: Longitudinal Failure, Transverse Failure, or Shear Failure. Because the failure index is a minimal ratio of stresses, the failure load can be computed by simply dividing the applied load by the failure index. Consider a transversely isotropic material whose principal axes are denoted as 1, 2. and 3, respectively, in which axis 1 is typically aligned with the direction of fibers. Particularly the transverse isotropy in a general 3D problem, there is a degree of arbitrariness in the orientations of axes 2 and 3, provided that they are together perpendicular to axis 1 and to each other. When any of the above stress ratios exceeds 1, failure is calculated, and the failure mode is found with the failure surface associated with the action plane of the respective stress component and the sense of the failure, i.e., tensile or compressive in the case of direct stresses, with the sense of the individual stress component; although experiments recommend that the actual fracture surfaces for compressive failure in general and oblique shear failure for brittle materials tend to fluctuate from the action planes of the respective stresses.

The maximum stress and strain criteria is indistinguishable in terms of their shear components since any shear stress or strain is decoupled from other shear, as well as any direct stress or strain, given the transverse isotropy of the material. To combine these two criteria, one only requires doing so for direct stresses and strains. With the conventional maximum stress criterion, by choosing different coordinate systems, one can obtain infinite numbers of different predictions for transverse failure.

Consider a pure shear stress state in the transverse plane. Through rotating the coordinate system about axis 1, the transverse stresses will fluctuate after the corresponding coordinate transformation rule. Deprived of exhausting all intermediate positions, two characteristic representations of the stress state are absolute shear (in the original coordinate system) and equal biaxial tension and compression (in the coordinate system rotated 45°).

Figure 5.7 depicts the result of progressive failure. Figure 5.8 shows the first ply failure survey. Results of the first ply failure stresses in +X-Direction: Longitudinal Failure in Ply: 1 is 1.23786E+07 (Pa), –X-Direction: Longitudinal Failure in Ply: 1 is –1.29168E+07 (Pa), +Y-Direction: Transverse Failure in Ply: 1 is 3.82122E+07 (Pa), –Y-Direction: Transverse Failure in Ply: 1 is –1.10690E+08 (Pa), XY-Direction: Shear Failure in Ply: 2 is 1.99134E+07 (Pa).

5.5.2 Tsai-Wu failure

The Tsai-Wu failure theory is employed on the top surface of 0° ply (Figure 5.9). Failure location is at the middle, and Tsai-Wu failure criterion is considered. Tsai-Wu utilized the failure theory to a lamina in plane stress. A lamina is failed if the failure equation is violated and is applied for orthotropic materials in their materials' principal axes. This failure theory is more familiar than the Tsai-Hill failure theory since it differentiates amongst the compressive and tensile strengths of a lamina. They contended that the failure envelope must be an ellipsoid and therefore stays closed. Autodesk progressive failure analysis software is utilized to predict failure stages of composite materials. Degradation factors are considered as 1, 0.01, 0.1,

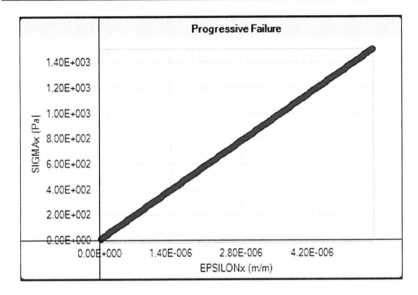

Figure 5.7 Progressive failure.

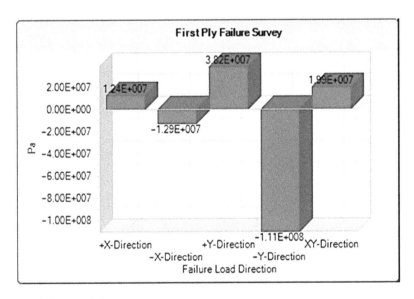

Figure 5.8 First ply failure survey.

and 1 with the failure criteria as f*: –3.00000E-01. Parameters selected are as follows. Delta Moisture (%): 2.60000E+01, Delta Temperature (C): 3.00000E+01, Stress Increment (Pa): 1.79400E+05, Y-Axis Stress – S22, X-Axis Stress – S11. Two hundred points are considered for this analysis, and sample stresses are provided in Table 5.3.

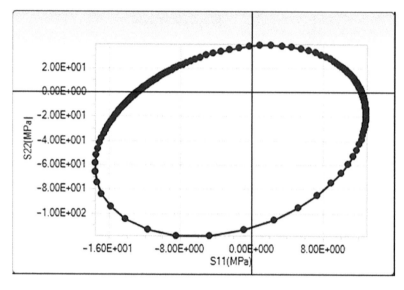

Figure 5.9 Elliptic failure loci determined to four anchor points on the axes.

Table 5.3 Sample stresses in X and Y axis

Step	S11	S22
1	1.23553E+01	6.98492E-16
2	1.23330E+01	3.89529E-01
3	1.23110E+01	7.78442E-01
4	1.22874E+01	1.16737E+00
5	1.22640E+01	1.55717E+00
6	1.22406E+01	1.94863E+00
7	1.22153E+01	2.34218E+00
8	1.21895E+01	2.73883E+00
9	1.21613E+01	3.13886E+00
10	1.21336E+01	3.54384E+00
11	1.21061E+01	3.95464E+00
12	1.20747E+01	4.37086E+00
13	1.20440E+01	4.79499E+00
14	1.20117E+01	5.22722E+00
15	1.19769E+01	5.66827E+00

Puck and Schürmann created the Hashin failure criteria. The fiber failure was dependent on material properties of the fiber in lieu of the properties of the ply, and the inter-fiber failure was separated into three including the transverse tension, moderate transverse compression, and large transverse compression. In addition, an equation was suggested to ascertain the angle of the fracture plane. Both the Hashin and the Puck and Schürmann criteria were 2D criteria that disregarded interlaminar stresses. Puck later revised his criteria to incorporate interlaminar stresses.

Fiber-reinforced laminated composites can reveal substantial stress reflections marked on three divergent geometric scales. These occur between the reinforcing fibers and yielding matrix material, which trigger further damage and a reduction in stiffness. There are stress factors at the interface between dissimilar composite plies in the ply level. Owing to the distinct Poisson's ratios of fiber and matrix, the fiber-perpendicular stresses yield internal stresses in the fiber direction.

Depending on the loads applied, the fiber's parallax tension or compression may increase or decrease. Laminated composite structures start to become damaged at loads that are far beneath the ultimate load of the structure. The recent failure criteria was termed as action-plane failure criteria, wherein the inter-fiber failure was computed centered on stresses that act on planes parallel to the fiber and inclined at an angle θ about the thickness direction. Once the damage is found to originate by the failure criteria, the material properties are degraded to replicate the cracks in the material. To validate the Puck damage prediction model, the progression of damage in a specimen must be examined. One method is to employ a non-destructive procedure that will allow for the recognition of damage progression in a composite structure.

The Puck fracture conditions for unidirectional fiber/polymer composites are broadly recognized based on the prediction of failure and post-failure degradation performance. Accordingly, the calculation of a laminated composite structure's ultimate load requires a Puck Failure Analysis wherein damage progression and associated stiffness reduction are recorded in the analysis.

5.6 MICROSTRUCTURE EFFECTS

Calotropis fiber composite fracture surface morphologies of the interfacial properties, such as fiber-matrix interaction, resin matrix fracture and loss, crack propagation and fiber pull-out of fibers from the matrix, were examined using a scanning electron microscope (SEM). It can be expected that the microstructural changes causing the in-plane tensile properties will not be altered, or only to a very small extent, by the presence of notches on the pin surface [12]. The SEM micrographs observed at the compression site show the fracture surfaces of the composite (Figure 5.10a, b), where debonding at the glass-Calotropis fiber interface is found. The space between the matrix and the fiber is generally visible as it is beginning to debond from the matrix, as depicted in Figure 5.10d. The cause for this may again be assigned to the expansion of a population of fiber defects and fiber ends with enhanced fiber content.

Figure 5.10 Fracture surfaces of chemically treated Calotropis fiber-reinforced composites.

The fiber defects (i.e., kinks) could act as a source of stress concentration in composites, as noted by the SEM micrograph (see Figure 5.10c). Flexural strength of the treated fiber composites was higher than that of untreated fiber composites. Including all the treated fiber composites, alkali-treated fiber reinforced composites had the highest flexural strength of all fiber contents. This could be related to the improved interfacial bonding.

The fiber breakage and matrix failure were detected, and it is possible that fiber treatment led to micro-fiber fibrillation where the surfaces become softer compared to untreated composites. In addition, this fibrillation could have spread over the actual surface area with the matrix in the composites, which may cause stronger fiber-matrix boundary adhesion and increased stress transfer. Improvement in mechanical properties is thus observed. The voids related to fiber pullouts, fiber breakage, matrix breakage, and fiber-matrix debonding is identified. The fiber is pulled out from the matrix due to poor adhesion, resulting in the formation of voids. Cracks develop close to the void areas because of continuous bending load on the samples. The cracks start where the applied load acts on the samples, leading to total failure of the composites. By enhancing the compatibility between the fiber and epoxy matrix, flexural strength may be improved.

5.7 CONCLUSION

The importance of novel materials and technology has been constant for years. The scientific community has performed accurate detailed studies about the increase in fundamental knowledge about natural fibers. The expansion towards low-cost and robust materials from renewable resources impacts sustainable growth. A desert plant as a dominant element in sandy soil and salty wadi beds which are rich in fibers was chosen for this study.

Researchers, engineers, and scientists have recently turned to natural fibers as a substitute reinforcement for fiber-reinforced polymer (FRP) composites. FRPs are extensively employed in the aviation, naval, and automotive industries for components that require a high ratio of strength to weight and durability. Hybrid composites are designed and fabricated using raw Calotropis procera fiber/glass with varying fiber weight percent as per ASTM standard.

The tensile strength, impact strength, density, and hardness of hybrid Calotropis procera fiber-reinforced composites have been assessed as new natural materials with superior properties. The growth in fiber content progressively increases tensile strength, impact strength, and hardness. The highest value of tensile strength for composites containing 30 gm Calotropis gigantea stem fiber was achieved with addition of 20 gm glass fiber. The above comparison clearly shows the usage of less than 20 gm glass fiber, which in turn increases tensile strength, impact, density, and hardness. The Izod impact strength was significantly reduced to 3.321 J when the glass fiber content was decreased. The resulting properties revealed that composites with good strength could be successfully developed using Calotropis procera fiber.

It was observed that the surface is smooth with lesser void content as shown on the upper surface of the composite sample. This lesser void is caused by an excess of resin squeezed during solidification, which causes a shortage of resins between two adjacent fibers. Despite the fact this study introduced a new natural fiber into the research, this study considered the weight proportion ratio of fiber to composite by 50% each, hence in the future this can be extended with various compositions using glass fibers.

The mechanical properties of these composite materials based on their properties were studied respectively. Using SEM (Scanning Electron Microscopy) analysis images of the breaking area for each sample that underwent a tensile test—by applying FTIR (Fourier Transform Infrared Spectroscopy) and EDS (Energy Dispersive Spectroscopy) analyses—spectrum bands were determined. Based on the results, it is recommended that these composite materials could be used for many lightweight applications.

REFERENCES

1. Pal, R. 2007. *Rheology of Particulate Dispersions and Composites*, Boca Raton: CRC Press.
2. Pal, R. 2005. New models for effective Young's modulus of particulate composites. *Composites B* 36: 513–523.
3. Velusamy, K., P. Navaneethakrishnan, and G. Rajeshkumar. 2018. The influence of fiber content and length on mechanical and water absorption properties of Calotropis Gigantea fiber reinforced epoxy composites. *Journal of Industrial Textiles* 48 (8): 1274–1290.
4. Velusamy, K., P. Navaneethakrishnan, and V. S. Arungalai. 2014. Experimental investigations to evaluate the mechanical properties and behavior of raw and alkali treated King's Crown (Calotropis Gigantea) fiber to be employed for fabricating Fiber composite. *Applied Mechanics and Materials* 598: 73–77.
5. Aruna, M. 2014. Mechanical behaviour of sisal fibre reinforced cement composites. *International Journal of Mechanical, Industrial Science and Engineering* 8(1): 84–87.
6. Roe, P. J. and M. P. Ansell. 1985. Jute reinforced polyester composites. *Journal of Material Science* 20: 4015–4020.
7. Khan, J. A., M. A. Khan, and M. R. Islam. 2014. A study on mechanical, thermal and environmental degradation characteristics of N, dimethylaniline treated jute fabric-reinforced polypropylene composites. *Fibers and Polymers* 15(4): 823–830.
8. Kim, H., Yutaka Miura, and Christopher W. Macosko. 2010. Graphene/polyurethane nanocomposites for improved gas barrier and electrical conductivity. *Chemistry of Materials* 22 (11): 3441–3450.
9. Dilli Babu, G., K. Sivaji Babu, and P. Nanda Kishore. 2014. Tensile and wear behavior of calotropis gigentea fruit fiber reinforced polyester composites. *Procedia Engineering* 97: 531–535.
10. Gandini, A., M. N. Belgacem. 2008. The state of the art. In *Monomers, Polymers and Composites from Renewable Resources*. Belgacem. Elsevier: Amsterdam. 1–16.
11. Aruna, M., S. Arivukkarasan, and V. Dhanalakshmi. 2014. Assessment of mechanical behavior of hybrid sisal reinforced composites. *Pensee Journal* 76 (4): 27–32.

12. ASM Handbook, Vol. 21, *Composites*, ASM International: Materials Park, OH, 2001.
13. Knopp, A. and G. Scharr. 2020. Tensile properties of Z-Pin reinforced laminates with circumferentially notched Z-Pins. *Journal of Composite Science* 4:78.
14. Lopez, G. A., M. Barrado, J. San Juan, and M. L. No. 2009. Mechanical spectroscopy measurements on sma high-damping composites. *Materials Science and Engineering* A521–2: 359–362.

Chapter 6

Green manufacturing and environment

Ali Sohani
K. N. Toosi University of Technology, Tehran, Iran

Jaroslav Vrchota
University of South Bohemia in Ceske Budejovice, Ceske Budejovice,
Czech Republic

Burçin Atilgan Türkmen
Bilecik Şeyh Edebali University, Bilecik, Turkey

Mohammad Mehdi Hosseini
K. N. Toosi University of Technology, Tehran, Iran

Hitesh Panchal
Government Engineering College, Gandhinagar, India

CONTENTS

DOI: 10.1201/9781003252108-6

6.1 INTRODUCTION

Green manufacturing (GM), which is gaining traction as an efficient and eco-friendly alternative to conventional manufacturing, is a solution to many problems facing mankind today. Green manufacturing can help address environmental protection, conservation, regulatory compliance, recycling, pollution control, waste management, and other related issues. As a result of optimizing resource use and reducing waste and pollution, GM can ensure a benign, harm-free product life cycle from conception to disposal, causing no or minimal adverse environmental impact. Reducing, reusing, recycling, and remanufacturing (4Rs) are fast becoming the model of growth and sustainable development around the globe [1].

Green manufacturing is considered in two ways:

1. Creating products environmentally friendly such as those used in renewable energy systems or clean technology machinery.
2. Environmentally friendly manufacturing that decreases pollution and waste by recycling, reducing emissions, as well as reduction of the use of natural resources.

Manufacturing with this concept goes beyond just addressing the environmental and social impact of pollution-centric processes [2]. Green production deals with process redundancy, ergonomics, and cost consequences due to incorrect production methods. Being faster or cheaper are no longer the only two criteria in which to assess a process line or the way a product is manufactured [3]. A manufacturing process is evaluated according to many factors, including materials used in manufacture, waste generation, effluents and their treatment, the lifespan of the product, and the treatment for the product after its useful life [4]. There is a great potential to replace high-carbon materials with those that have less carbon [5]. Buildings, for example, can generally use timber or pozzolan-based concrete as a substitute for Portland cement to help mitigate pollution. Moreover, there is also the possibility to improve the quality of the materials' processing system. Companies can now incorporate green packaging into their products in addition to using greener materials, such as innovative bio-based materials—wheat straw packaging is one example that can reduce production water by 90% and energy consumption by 40% [6].

6.2 RECYCLING AND REUSE

Recycling is a process in which waste materials are transformed into new materials. One important part of this concept is the energy recovered from waste materials. The benefit of a material that contributes to regaining its original particles and properties is called recyclability [7]. Recycling not only prevents the waste of potentially useful materials, but it can also reduce fresh raw material consumption, energy consumption, as well as air and water pollution. As part of the "Reduce, Reuse, and Recycle" waste hierarchy, recycling is a vital component of modern waste reduction [8]. Here are some advantages of recycling:

- Helps conserve valuable natural resources
- Reduces the amount of trash sent to landfills
- Saves energy and money
- Creates new jobs
- Reduces greenhouse gas emissions

Recyclable products reduce raw material input and redirect waste output in the economic system, thereby promoting environmental sustainability [9]. Paper, cardboard, glass, metal, plastic, tires, textiles, batteries, and electronic devices are just a few of the materials that can be recycled. Biodegradable waste is also recycled through composting and other means, such as by using food waste or by gardening [10]. Materials from household recycling bins or curbside recycling bins are collected and processed.

When done correctly, a material can be recycled into a new material—for example, used office paper can be repurposed as new office paper. It is possible to remanufacture certain materials, such as metal cans, repeatedly without affecting their purity and quality. Salvaging material components from complex products is another method for recycling, depending on whether the constituents are valuable or not.

Generally, the operation of recycling is divided into two main types: internal and external. In internal recycling, waste materials and particles produced in a manufacturing process are reused. This kind of recycling is widespread in the metal industry. Take the copper industry as an example: The production of copper pipes causes a certain amount of waste in the form of the end of the pipe and cutting, which are remelted and recast. In external recycling, old materials are reclaimed from products no longer in use or that have become outdated. For example, newspaper and magazine scraps can be collected for re-pulping and then manufactured into new products.

Some examples of everyday objects externally recycled on a large scale are aluminum cans and glass bottles. There are three ways through which these materials can be amassed: The first is drop-off centers, where consumers leave waste materials but do not receive any money in return; curbside collection, where waste materials are sorted and left by the curb to be collected by an agency; and buy-back centers, where waste materials that consumers sort and bring in are purchased [11] (Figure 6.1).

The definition of reuse is utilizing an item more than once, whether for its main objective (conventional reuse) or to accomplish a disparate operation. Reuse is considered a sensible way to save resources, money, energy, and

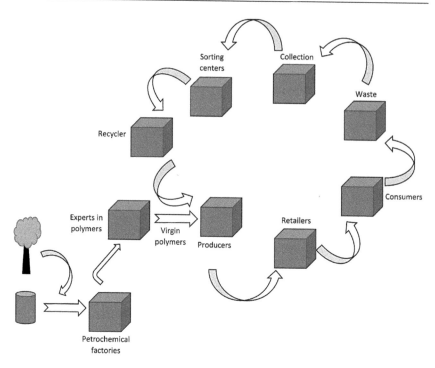

Figure 6.1 Product life-cycle stages.

time. From an economic viewpoint, many jobs would be created by this service, and it may also provide business activity that could be beneficial to economic growth while giving organizations and those with low incomes easy access to high-quality products [12]. The usual method of delivering milk in glass bottles is an example of conventional recycling; other examples are tire retreading and using recyclable plastic boxes as opposed to single-use corrugated fiberboard boxes as shipping containers.

Formal life-cycle assessment is often the best way to determine how the various effects of reuse interact. In comparison to the full product life cycle, researchers have found that with the process of reusing, carbon dioxide emissions are reduced by more than 50%. Thus, utilizing reused products is an effective method of reducing carbon dioxide emissions and a carbon footprint. In many cases, manufacturing and the supply chain have unknown carbon footprints.

6.3 COMPARATIVE ANALYSIS WITH DIFFERENT MATERIALS

In this next section, in order to provide the possibility of comparison, the recycling processes of aluminum, paper, and plastic—the three widely used materials in the industry—are described, respectively.

6.3.1 Aluminum

Aluminum could be recycled using the process found in Figure 6.2. Aluminum recycling provides many advantages [13]:

- Compared to producing aluminum from raw materials, recycling aluminum uses 95% less energy.
- In addition, 97% of the greenhouse gas emissions produced in primary production are diminished.
- Approximately nine tons of CO_2 emissions are saved for every ton of aluminum recycled. Four tons of bauxite, which is used to produce aluminum, are saved. One thousand kilograms of CO_2 are equal to driving over 5600 km.
- There is a permanent supply of the metal. The material is totally recyclable forever, so every time it enters the recycling loop, carbon emissions diminish.
- It is 20 times more energy efficient to produce aluminum cans from recycled metal instead of utilizing primitive metal.
- After recycling empty drink cans, brand-new cans can be sold within 60 days.

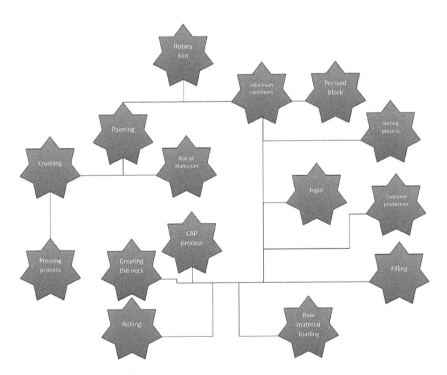

Figure 6.2 Recycling of aluminum.

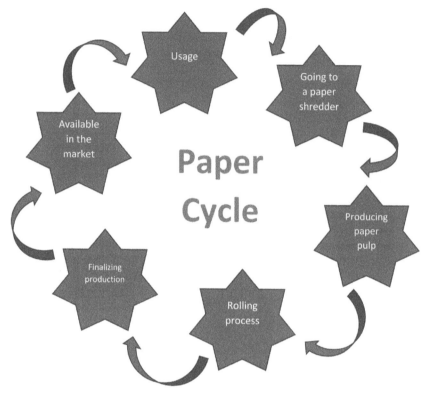

Figure 6.3 Recycling of paper.

6.3.2 Paper

The recycling process of paper is shown in Figure 6.3. Paper is made from about one-third of all trees that are cut down, which is 160,000 km² of forest logged each year. Despite the fact that trees are considered renewable resources, only 9% of the trees used to produce paper are indigenous to old forests. When paper is recycled, fewer trees are planted as a result. In addition, although recycling paper saves 40% of the energy it takes to make it from scratch at the same time, modern paper mills use waste wood, while recycling plants use fossil fuels—so that is the case against. Factors in favor of recycling are harmful compounds utilized by paper mills like methanol, toluene, and formaldehyde. A report published by the US Environmental Protection Agency illustrates that paper mills are among the industries that produce the highest amount of pollution in the US. In comparison with making new paper, recycling reduces water pollution and air pollution by 35% and 74%, respectively. Moreover, 3m³ of landfill is omitted by recycling 1000 kg of newspaper. Methane, a powerful greenhouse gas, is generated as a result of paper decomposition in the ground. Recycling paper generally seems to be much more environmentally friendly than making it from fresh pulp [14].

6.3.3 Plastic

Figure 6.4 introduces the recycling process of plastic. Diminishing plastic waste is a crucial issue for both humans and the Earth. The advent of new technologies has provided a multiple solutions such as utilizing chemical processes to transform after-consumption plastics into fossil fuel replacements and decomposing plastics that could not previously be recycled. These kinds of technologies can benefit the environment [15].

- Energy conservation: Recycling 1000 kg of plastic saves more than 5kWh of energy, which is equal to the amount of energy consumption of two people in a year.
- Decreasing the use of petroleum: Approximations say that by recycling plastic waste, up to 40% of oil consumption could be lessened. That is 16.3 barrels of oil per 1000 kg of recycled plastic.
- Reduction in CO_2 emissions: Reduction of oil consumption results in the reduction of CO_2 emissions and other greenhouse gases while new plastics are being produced.
- Reduction in the use of landfills: Fewer plastics in landfills equals fewer emissions of common landfill gases like CH_4 and CO_2. Both these gases affect the air and the ground, resulting in environmental detriment and public health problems.

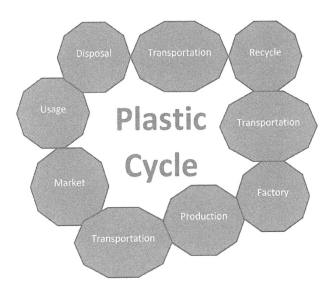

Figure 6.4 Recycling of plastic.

6.4 LIFE-CYCLE ASSESSMENT

6.4.1 General description

Green manufacturing refers to the system of designing manufacturing processes to prioritize energy conservation, environmental sustainability, and improved health and safety for communities, employees, and consumers. Life-cycle assessment (LCA) is one of the primary tools used to assess the environmental sustainability of a product, process, or service by identifying and quantifying material and energy usage—as well as waste output—at each stage of its life [16]. Every product or process has different phases or stages in its life, each of which is made up of a variety of activities. These stages are broadly defined for industrial products as raw material extraction, manufacturing, use and maintenance, and end-of-life—also including all transport and waste management in the system (Figure 6.5) [17].

According to the International Organization for Standardization (ISO) 14040 and 14044 standards [18, 19], there are four phases in an LCA: goal and scope definition, inventory analysis, impact assessment, and interpretation of the results. Figure 6.6 presents the LCA framework. The goal and scope definition stage defines the purpose of the study, the functional unit,

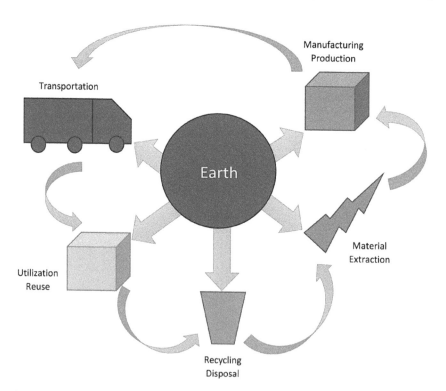

Figure 6.5 Life-cycle assessment.

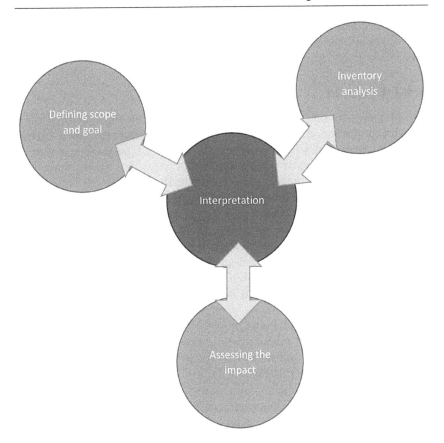

Figure 6.6 Life-cycle assessment framework.

the system boundaries, and the assumptions for the study on which the evaluation is based. The second step of the LCA is inventory analysis, which identifies the processes required for raw materials production; use of auxiliary materials, water, and energy during manufacturing; emissions to air, water, and soil during manufacturing; transport of raw materials; and packaging materials. An impact assessment calculates the product's long-term effects on human health, ecosystems, and natural resource depletion based on inventory data. The study's conclusions and recommendations are then presented in the interpretation step. Further evaluation—such as validation, comparison with other studies, and sensitivity analysis—can be included at this step [20].

LCA is widely used as a tool to determine the environmental impacts and to support policies and performance-based regulation, particularly in the field of manufacturing. Several life-cycle assessment studies have been conducted over the years to evaluate the environmental sustainability of a wide range of products from different sectors, including energy [21–24], building and construction [25–27], food and agriculture [28–30], and transportation [31, 32]. Some studies cover the entire life cycle [33, 34], while others focus on a single stage [35, 36] or process [37, 38].

6.4.2 Example

The previous sections illustrated an introduction to the topic, description of recycling and reuse, and comparative analysis of different materials. Having provided a general description in the previous part of this section, in the current part of this section we present a case study related to the environmental impact assessment of a product to illustrate further how the life-cycle assessment can be used [39]. Glass is selected as a material, one that is inorganic, fragile, and hard. It has an amorphous structure and is commonly found in containers, mirrors, bottles, pots, home glass, and artwork.

The case study aims to estimate the life-cycle environmental impact of glass packaging used in the beverage industry. Life-cycle analysis has been applied according to the international standards ISO 14040 and ISO 14044 [18, 19]. One kilogram of formed and finished white beverage packaging is defined as a functional unit. As demonstrated in Figure 6.7, the system boundaries include the following life-cycle stages: facility construction, raw material extraction and manufacturing, raw material transportation, glass packaging production, and waste management as well as transportation.

The inventory data is presented in Table 6.1. The chosen facility is a typical glass manufacturing facility with natural gas-fired furnaces in operation. Batch preparation, preheating, oven, melting, annealing, shaping, and packaging steps are included in the glass beverage packaging production stage. The factory is built, and the factory's systems are installed during the facility installation stage. The main raw materials used in the production of glass beverage packaging are silica sand, soda ash, dolomite, limestone, and other raw materials. In the glass industry, cullet is used as a raw material.

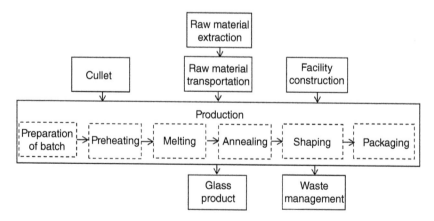

Figure 6.7 Life-cycle stages of glass packaging.

Table 6.1 Inventory data

Raw materials	kg/functional unit
Silica sand	0.35
Soda	0.11
Feldspar	0.04
Dolomite	0.02
Limestone	0.11
Cullet	0.50
Transportation	**Truck/km**
Silica sand	40
Soda	785
Dolomite	40
Limestone	40
Cullet (external)	100
Packaging material	150
Energy	**MJ/functional unit**
Electricity	0.92
Natural gas	3.75

The CML 2001 method [40] has been selected for the analysis of the environmental impact categories. The following impacts are considered: abiotic depletion potential elements (ADP), abiotic depletion potential fossil (ADP fossil), acidification potential (AP), eutrophication potential (EP), freshwater aquatic ecotoxicity potential (FAETP), global warming potential (GWP), human toxicity potential (HTP), marine aquatic ecotoxicity potential (MAETP), ozone layer depletion potential (ODP), photochemical ozone creation potential (POCP), and terrestrial ecotoxicity potential (TETP).

The environmental impacts are presented in Figure 6.8. The total GWP per kilogram of glass bottle is estimated at 1.2 kg CO_2 eq. The production stage accounts for 84% of total GWP, followed by raw materials extraction and processing (7%), transportation (5%), plant installation (3%), and waste management (2%). Because of its high energy consumption, the glass production step has the greatest environmental impact (4.6% ADP–89.0% MAETP). Improvements in energy and raw material inputs are required to reduce emissions from the manufacturing step as a result of the environmental sustainability assessment.

6.5 CASE STUDY: GREEN MANUFACTURING IN THE CZECH REPUBLIC

The top priority of most manufacturing companies will no longer be to increase productivity, but instead to approach sustainability and social

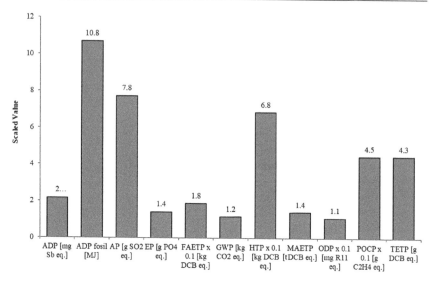

Figure 6.8 Impacts of glass packaging.

responsibility. Can we expect that increasing productivity of companies will help reduce unemployment or vice versa? The current industrial revolution, also referred to as Industry 4.0, is expected to grow exponentially from both technical and socioeconomic perspectives.

However, sufficient sustainability requires a combination of technological and social innovation. Among the biggest social problems is the risk of job losses due to the high automation of operations. In turn, new job opportunities may arise in high-skilled positions. Businesses should pursue sustainable development, i.e., good economic performance and social benefits, e.g., increased access to education and retraining. Industry 4.0, as well as Society 5.0, has great potential to achieve positive impacts on the economy as well [41].

For an organization to succeed in this environment, attention needs to be paid to training, learning, and innovation capabilities. It is for this reason, considering the literature [42–50], this model for green manufacturing has been developed, supported from the Industry 4.0 and Society 5.0 directions, which are primarily saturated with Environmental, Technological, Social, and Economic aspects (Figure 6.9).

6.5.1 Conditions of the Czech Republic

As mentioned at the beginning of the chapter, the countries with a strong industrial base are paying the most attention to the advent of Industry 4.0. The Czech Republic is one of these countries, whose share of industry in gross value added is approximately 32%. The Czech Republic thus has a commitment to the population, which is that we should be among the first

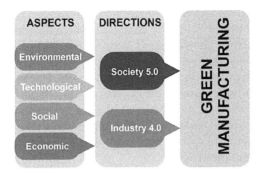

Figure 6.9 Green manufacturing framework.

in the processes set in motion by Industry 4.0, as the historically high share of industry in GDP makes us predisposed to this step in the context of sustainability. To sustain it, the Czech Republic must be prepared to play the role of a capable, cooperative partner that can integrate new technologies quickly and—thanks to a high level of industrial innovation—maintain itself among the region's leaders. In particular, legislation and infrastructure must be adapted to this [51, 52].

For the above reasons, the Ministry of Industry and Trade issued the strategic document National Initiative Industry 4.0 in September 2015. This document is conceptually followed by the Industry 4.0 Initiative approved by the government on August 24, 2016. This initiative aims to mobilize key sectors to develop detailed action plans to compete in the new industrial phase [52, 53].

A number of experts from the academic, public, and private spheres, led by Prof. Mařík [54] participated in the preparation of the comprehensive document, National Initiative Industry 4.0. The aim of the document is to prepare key production ministries and to connect the business and social spheres with regard to sustainable development in line with the environment.

The document consists of 11 chapters that provide global information, complemented by a SWOT analysis, which shows that the biggest threats associated with Industry 4.0 in our conditions include the late or poor-quality implementation of the basic digital communication infrastructure necessary for the Industry 4.0 concept and the lack of standardization for IoS, IoT in conjunction with environmental protection. A real threat, which is already historically determined, may be the ever-deepening dependence on Germany, as the direct export of industrial production to Germany is almost 30%. And if we take into account the links with German companies operating in the Czech Republic, we depend on Germany for at least half of our industrial production.

On the other hand, a major opportunity for the Czech Republic may be to increase its attractiveness to new foreign investors thanks to appropriate investment incentives and an adequate workforce [52, 55]. As the initiative implies, this will not only involve changes in the industrial sector, but also social changes, which should mainly concern the field of education, as the Czech Republic is permanently facing a significant problem in the form of a shortage of skilled labor in industrial production. The initiative primarily refers to the mismatch between the existence of job opportunities in technical fields on the one hand and the focus of most university graduates on the humanities on the other. Secondarily, this implies the need to adapt the entire education system to the upcoming changes that will place more emphasis on an interdisciplinary approach.

6.5.2 Parametric considerations

For the case studies, enterprises were selected that belong to the top 10% according to Vpi4 and also represent enterprises by size (small enterprise [<50 employees], medium enterprise [<250 employees], large enterprise [>250 employees]), thus fulfilling the idea of the best enterprises that have the introduction of new technologies and environmental and sustainability considerations in line. In the case studies, the focus was on the modernization of production in recent years with regard to Industry 4.0 and social and environmental aspects. At the same time, company representatives were asked about human resource-related barriers.

Case study A

a. Company information

This is a manufacturing company that operates in the domestic and foreign markets in the field of electronics and electrical engineering. The largest share of the company's revenue consists of electrical products for the automotive industry. The company with more than 40 years of tradition currently emphasizes the development of its own products with high added value, and in the field of custom electronics production, increasing the quality and efficiency of the production process, including its own development of special single-purpose production machines with the aim of technological mastery of demanding tasks.

The company currently employs 310 employees and in 2021 reported earnings of approximately CZK 1 billion. According to one of the board members, the company has often used subsidy incentives for SMEs for innovation and modernization in the past, thanks to which it has managed to grow and keep up with the competition. Thanks to acquisitions, it had more than 250 employees the year before last and lost these opportunities. The company has so far implemented an MES system (xTrace) and some sub-areas as part of its digital transformation. It has also acquired lasers, labelers, and scanners to tag all

---

incoming material with QR codes with tracking throughout the production process. However, the implementation to date has not yet ensured machine-to-system and system-to-system communication, which the owner says is currently happening. "The priority for us is the introduction of two-way machine communication, for example, automatic program selection on the production machines. Furthermore, the communication link between the ERP system and the MES system (xTrace) will be completed. To ensure the integration of the acquired technology, an upgrade of the ERP system will be implemented, which will be supplemented by production plan visualization boards at the production sites. An interface will be implemented for two-way communication between the system elements via Ethernet and Wi-Fi. The tracking of work in progress in transport units (gulls) will also be implemented".

b. Past status

The goal was to modernize the MES. The old MES system implementation was based mostly on data collection and basic functionality of the production workstations. It lacked tool and fixture management and data visualization. It also lacked a quality management module—for example, an emergency brake system or automatic line stops when quality control was required. "Previously, there was an hourly board from the MES at each workstation tracking the performance and status of the workstation." The status of jobs could be tracked in the MES or ERP, but the system did not provide a clear and configurable output.

Previously, tools, machines, and products were identified using QR codes, with manual assistance (code reading) by the operator. At the same time, the communication of the individual elements of production was only one-way. The traceability of material and work in progress was by means of marking the material with QR codes by laser or labels, then the flow of material through the production process was monitored with the manual assistance of operators. Quality control was also an issue, which was carried out visually and on electro-testers.

c. Current status

The company has just completed the integration of the MES at a cost of almost CZK 7 million, where two-way communication with the machines has been achieved, i.e., automatic selection of process data on the machine without operator intervention. Furthermore, it has introduced two-way communication between other elements of the system (ERP control software of the assembly lines). The quality control module includes first piece inspection control and periodic inspection control. This control includes automatic stopping of production lines during quality checks and automatic evaluation of the results.

An important module is the emergency brake function. Tool and fixture management is also integrated. This add-on manages the stock of tools and jigs, their wear (cycles used), and, last but not least, their correct use, independently of the operator. The introduction of Supermarket logic, intermediate stores, and RACK logic is also a benefit.

In the framework of this year's project, solar panels were put into operation, and more powerful batteries were installed, which currently cover 30% of the electricity consumption for daily production, as well as RFID tool identification with automatic online communication with the machines, with automatic control of the tool location. There has been an upgrade of the systems to support RFID tool identification and the implementation of the "MES Interface" for two-way communication between the system elements over Ethernet and Wi-Fi. The interface allows bi-directional communication between the MES and the work machine. "The laser tags the material with QR codes. The machine requests the serial numbers of the incoming material, which are checked against the data provided by the MES. Once tagged, it scans the QR codes and sends them back to the MES".

The system allows the company to find out the current location of material within production and especially work in progress. This helps planners and other production staff to more easily compare the real plan and identify potential delays and other deviations from the planned status. At the same time, transport units with work in progress are identified by barcodes.

The loading of products into individual transport units was introduced. The transport units are identified by barcodes, and handling points are equipped with optical readers. "As a result, the tracking of semi-finished products within the crates, with the tracking of their flow in production, has led to a significant improvement in work-in-progress status." A material scanner with automatic supplier and material recognition was also acquired, followed by 2D labeling of the material and automatic verification of correct labeling. This ensures that all material is correctly labeled for traceability, eliminating human error. The investment in this system alone was around CZK 2 million.

In addition, a system using artificial intelligence for learning to recognize visual defects using inspection cameras was installed at three workstations. This was achieved through the purchase of three industrial scanning cameras with fixed lenses (approx. CZK 250 thousand), enabling measurement of shapes and dimensions, detection of surface defects in materials, inspection of electronic components, reading of barcodes and QR codes, and purchase of the necessary software for automatic defect recognition.

d. Barriers in relation to the social, environmental, and economic pillars

Bureaucracy and lack of qualified personnel were among the problems noted by one of the owners during the modernization of production, while the factor of constant innovation in the PV cell and battery poly and related projects that had to be updated regularly were also problematic. The final modernization, he said, would mean that: "Some positions can be replaced by automation processes, which will reduce the human error rate, while new demands will be placed on the staff performing the control function. The advantage here is that some existing staff already know how to do this—it is just a matter of retaining them and convincing them not to go to Bosch, for example. At the same time, the first steps have been taken to make the company independent in terms of energy supply."

Case study B

a. Company information

This is a company whose main business is the manufacture of consumer electronics, where its primary market is the large venue audio segment (cinemas, theatres, auditoriums). The company was founded 12 years ago with the aim of developing and manufacturing exceptional professional sound systems of its own design. In the words of the owner, "Our loudspeaker systems, amplifiers, and sound systems are characterized by a different design approach to sound reproduction compared to the main competitors. Our aim and goal is to create a complete electroacoustic chain in which there is no distortion of sound, from the signal source to the listener. This is in contrast to competitors who focus on all sorts of digital correction of unwanted distortion and sound modification".

The company posted a revenue of over 250 million crowns last year with a total of 85 employees. The company's revenue has grown regularly at about 15%, mostly due to frequent upgrades. At the same time, the company does not focus on the domestic market, which accounts for 10% of revenue, but mainly foreign markets. Over the past few years, the company has invested primarily in ERP upgrades, 3D printing, data storage, and the cloud.

b. Past performance

The company used an information system that covered all the company's main activities: "It included production planning and management, sales, purchasing, warehouse management, and financial management. The information system could be accessed via a character terminal interface, a thin graphical client based on Microsoft .NET technology and via a scaled character interface on radio frequency terminals." There was no BI or MIS integrated in the ERP. There was also a problem with a high percentage of scrap and waste during new product development, or the production of specialized, bespoke production, where machines were not able to work optimally with the material when producing atypical or innovative products.

There was also a problem with data and access to it. Users worked with data stored on their disks or on shared folders. This data was regularly backed up daily, but only the latest versions of documents from the previous day. It was also necessary to boot the user on each device he or she needed to work on or connect to. Huge amounts of sometimes sensitive data circulated around the company via email. "3D printing was a big question mark for the company, with the advantage of development on the one hand, and time and choice of devices that none of us had a clue whether to go for or not on the other."

c. Current status

The ERP has been upgraded at a cost of just under CZK 5 million, or the implementation of a new version of the ERP. The new version of the information system brought an adaptive and independent user environment that supports the personalization of each role, activity, and user. Thanks to HTML5 support, it is also possible to connect to the new user environment using mobile devices. ERP allows users to extend existing applications and create new ones that do not need to be reprogrammed during the next update of the information system. "IT sees the main benefit in increased security and application performance and improved ability to integrate with external applications by supporting new versions of ODBC and JDBC." It is possible to quickly customize any process through flexible visual process maps that can be used to create and modify processes. The IS also includes a BI module to provide an overview of the state of the business and assist in strategic decision-making. BI includes built-in flexible metrics, role-specific operational dashboards, and advanced analytics using a variety of data sources. These outputs are available on mobile devices.

Last year, a private cloud for administrative and production data was established (4x NAS server 2U, with storage for administrative and production data, total capacity 32TB + 64TB backup storage). The deployment of the corporate cloud created a central storage for user documents. Data from computers is automatically synchronized, and each document is stored in several versions on the backup storage. A central password and access management is set up in the cloud, user accesses are authenticated through this server, and there is no need to boot to each device.

"We have purchased a 3D printer that will allow us to expand the use of 3D printing both in product development (production of samples, prototypes, atypical products) and in the actual production of audio components." It is a 3D printer with a workspace of 25 x 21 x 21 cm and a 0.4 mm nozzle with layer height of 0.05 mm. Supported materials are PLA, ABS, PET, HIPS, Flex PP, Nylon, Bamboofill, carbon filament, and polycarbonate. It is operated by two employees and is used, according to the manager's estimate, once a week. Thanks to sophisticated 3D modeling and almost residue-free production of these prototypes, material consumption has been significantly reduced. They have invested an estimated 300–400 thousand crowns in 3D printing, including servers for cloud storage.

d. Barriers in relation to social, environmental, and economic pillars
"The biggest problem they initially saw was what systems to put in place and how to manage the whole project with regard to business continuity; however, it turned out that a much bigger problem was the unexpected loss of several long-time loyal employees, when both of them became long-term ill within two weeks. Fortunately, their absence was made up with two new recruits."

Case study C

a. Company information
Since its establishment in 1999, the company has developed and manufactured measuring instruments of its own design used for metering on distribution networks and local distribution systems. The development of measuring instruments involves its own team of employees with many years of experience in the power industry and a number of external collaborators/specialists. The company markets and supplies a range of instruments and specialized solutions for specific problems. Thanks to the potential and flexibility of the development department, the company can apply theoretical solutions to individual requirements in practice very quickly. Recently, in addition to measurement technology, the development department has been working on ground fault indications in HV networks, including the software superstructure. The company provides its own special software for all supplied instruments, expert operator training, subsequent data analyses, and consultation with proposed solutions resulting from the analyses. The company's main products are universal measuring instruments operating at all voltage levels for voltage, current,

power and energy evaluation, which can record daily diagrams on selected days and register voltage events (dips, increases, and interruptions).

"Everything we manufacture and ultimately sell is developed by us." The company sees investment in human resource development in the company as very substantial. It recognizes the competitive advantage that educated, skilled, and motivated employees bring. Innovation, maximum production quality, and customer orientation are the means to achieve the company's goals. Last year the company achieved earnings of CZK 180 million with 42 employees.

b. Past status

For years, the company has struggled with a high proportion of overhead in terms of energy requirements to maintain optimal production conditions. It invests a large part of its financial resources in developing and automating production, where in the past year, production planning was carried out from the QAD information system. The company carried out long-term, medium-term, and short-term planning, but only up to unlimited capacities; the information system could not plan up to limited capacities or carry out production planning according to customer priorities.

Orders were planned in QAD according to the production time of each operation, backwards according to the required completion date of the order. Long-term planning was generally done once a year when creating the annual sales plan. Data from the information system was converted into an EXCEL spreadsheet from which the projected annual financial sales volume was calculated, and the total capacity requirement of this plan was calculated and used to plan the annual production capacity.

For medium-term planning, QAD drew up a production master plan according to the orders and schedules received, which were then checked by the company through bottlenecks. This was used by the shop floor planners to draw up a short-term production plan for the selected production sites in an EXCEL spreadsheet. "But if we had a complication during the day that disrupted this daily workplace plan, it was always necessary to address it expeditiously and then verbally inform the production operator of the change."

c. Current status

The production hall has been upgraded with insulation and a closed-loop heat recovery system linked to heat pumps, which have been linked to the existing photovoltaic cells that had been installed, reducing energy costs by 80%. In addition, a system for production planning in a capacity-constrained environment was implemented to simplify and improve the management of the master production planner. "The system performs production plan balancing updates, where it compares process capacity with client and customer requirements according to ERP data and then exports the balanced production plan to the ERP, which updates other processes according to the plan." APS offers several optimization methods for plan optimization and can take into account priority work orders. The planning is done automatically.

According to the selected rules, the system optimizes the sequence of work order operations. The scheduler then solves "only" the jobs for which the system has not found the possibility to meet the deadline. For these jobs, the scheduler receives information on how many hours/days it needs to find to fulfill the job (e.g., machine capacity increase, cooperation, etc.). The availability of materials, semi-finished products, as well as the quantity of materials on purchase orders, is checked as part of the planning.

Central control for production lines has been introduced. The central control allows automatic uploading of production software from the control PC to the individual setup machines within the line. It is also possible to centrally manage any substitutions in the piece lists and to easily select variants of the production SW. In the words of the Chairman of the Board of Directors, "The system has helped a lot to improve the efficiency of the planning process and to introduce discipline in the processes related to advanced planning (checking out work done, receiving materials into stock, etc.), i.e., everything that is necessary to make the planning process work in a controlled and efficient manner". The implementation of the system, including visualizations of plans and line occupancy, cost the company approximately CZK 2.5 million.

d. Barriers in relation to human resources

"Quality employees." During the conversation on this topic, emphasis was placed on the fact that trained professionals are not in the market, school graduates do not know how to do it and need to be "trained" and then retained. The owner of the company feels proud of the talent-management system that is in place in the company, whereby after identifying suitable candidates, special attention and care is given to them. Thanks to this, they cultivate a social and ecological mindset, and many of them subsequently implement new sustainability systems in the company.

6.6 CONCLUSION

Every era brings with it different possibilities and opportunities, but it is up to businesses to decide how they approach these opportunities and whether they can use them to their advantage. Even the Fourth Industrial

Revolution has its advantages and disadvantages. However, if we can define them early, incorporate them, and adapt to them, we need not fear it. The aim of Industry 4.0 is to make production processes as efficient as possible and to interconnect the activities and services associated with them, with an emphasis on sustainability and the environment. More specific objectives include: reducing emissions, reducing costs, increasing labor productivity, improving quality, social responsibility, and making it easier to meet individual customer requirements. More than ever, it will be necessary to be flexible, with the ability to make quick judgements and, above all, not to give up on education, but rather to continue to actively increase knowledge and self-educate.

Green manufacturing is not something we can expect in a year or two; it will take at least 10 years to develop. This emerging era, full of modern technology, digitalization, and the Internet, will need ever greater numbers of skilled human resources—as the case study results also confirmed—but it opens an imaginary gateway for the next generation.

REFERENCES

1. M. A. Rehman, R. Shrivastava. Green manufacturing (GM): past, present and future (a state of art review). *World Review of Science, Technology and Sustainable Development*. 10 (2013) 17–55.
2. E. Saedpanah, R. Fardi Asrami, A. Sohani, H. Sayyaadi. Life cycle comparison of potential scenarios to achieve the foremost performance for an off-grid photovoltaic electrification system. *Journal of Cleaner Production*. 242 (2020) 118440.
3. A. Sohani, M. Zamani Pedram, K. Berenjkar, H. Sayyaadi, S. Hoseinzadeh, H. Kariman, et al. Techno-energy-enviro-economic multi-objective optimization to determine the best operating conditions for preparing toluene in an industrial setup. *Journal of Cleaner Production*. 313 (2021) 127887.
4. K. Balan. Introduction to green manufacturing. *The Shot Peener*. 22 (2008) 4–6.
5. A. Sohani, H. Sayyaadi. End-users' and policymakers' impacts on optimal characteristics of a dew-point cooler. *Applied Thermal Engineering*. 165 (2020) 114575.
6. J.-P. Tricoire. Here's why green manufacturing is crucial for a low-carbon future. *World Economic Forum*, 2019.
7. G. Villalba, M. Segarra, A. Fernandez, J. Chimenos, F. Espiell. A proposal for quantifying the recyclability of materials. *Resources, Conservation and Recycling*. 37 (2002) 39–53.
8. J. Lienig, H. Bruemmer. Recycling requirements and design for environmental compliance. *Fundamentals of Electronic Systems Design*. https://doi.org/10.1007/978-3-319-55840-0_7; (2017) 193–218.
9. M. Geissdoerfer, P. Savaget, N. M. Bocken, E. J. Hultink. The Circular Economy–A new sustainability paradigm? *Journal of Cleaner Production*. 143 (2017) 757–68.

10. M. Pamela, M. Christine, G. Mamatha. *The Garbage Primer.* ISBN 978-1558212503, 1993.
11. Britannica. Recycling. https://www.britannica.com/science/recycling; Accessed on October 15, 2021.
12. Waste disposal & recycling|waste management. Reuse www.wm.com; Accessed on October 16, 2021. 2021.
13. Alupro - The Aluminium Packaging Recycling Organisation. alupro.org.uk; Accessed on October 18, 2021. 2021.
14. I. Čabalová, F. Kačík, A. Geffert, D. Kačíková. *The Effects of Paper Recycling and Its Environmental Impact.* InTech, Rijeka 2011.
15. U.E. Team. *Plastic Recycling Technology: What are the Environmental Benefits?* UBQ, 2020.
16. A. Azapagic. Assessing environmental sustainability: Life cycle thinking and life cycle assessment. in: A. Azapagic, S. Perdan, (Eds.), *Sustainable Development in Practice: Case Studies for Engineers and Scientists.* Second ed. John Wiley & Sons, Ltd, Chichester, UK, 2010. pp. 5680.
17. J. B. Guinee, R. Heijungs, G. Huppes, A. Zamagni, P. Masoni, R. Buonamici, et al. Life cycle assessment: Past, present, and future. *Environmental Science & Technolog,* 45 (1) (2011) 90–96.
18. ISO. *Life Cycle Assessment - Principles and Framework.* International Standard Organization, Geneva, Switzerland, 2006. pp. 1–20.
19. ISO. *Life Cycle Assessment - Requirements and Guidelines.* International Standard Organization, Geneva, Switzerland, 2006. pp. 1–46.
20. H. Baumann, A.-M. Tillman. *The Hitch Hiker's Guide to LCA: An Orientation in Life Cycle Assessment Methodology and Application.* Studentlitteratur AB, Lund, Sweden, 2004.
21. B. Atilgan, A. Azapagic. Life cycle environmental impacts of electricity from fossil fuels in Turkey. *Journal of Cleaner Production.* 106 (2015) 555–64.
22. B. Atilgan, A. Azapagic. Renewable electricity in Turkey: Life cycle environmental impacts. *Renewable Energy.* 89 (2016) 649–57.
23. J. Cooper, L. Stamford, A. Azapagic. Environmental impacts of shale gas in the UK: Current situation and future scenarios. *Energy Technology.* 2 (2014) 1012–26.
24. C. Gaete-Morales, A. Gallego-Schmid, L. Stamford, A. Azapagic. Assessing the environmental sustainability of electricity generation in Chile. *Science of the Total Environment.* 636 (2018) 1155–70.
25. R. M. Cuéllar-Franca, A. Azapagic. Environmental impacts of the UK residential sector: Life cycle assessment of houses. *Building and Environment.* 54 (2012) 86–99.
26. B. Atılgan Türkmen, T. Budak Duhbacı, Ş. Karahan Özbilen. Environmental impact assessment of ceramic tile manufacturing: A case study in Turkey. *Clean Technologies and Environmental Policy.* 23 (2021) 1295–310.
27. M. Manjunatha, S. Preethi, H. G. Mounika, K. N. Niveditha. Life cycle assessment (LCA) of concrete prepared with sustainable cement-based materials. *Materials Today: Proceedings.* 47 (2021) 3637–44.
28. X. C. Schmidt Rivera, A. Azapagic. Life cycle environmental impacts of ready-made meals considering different cuisines and recipes. *Science of the Total Environment.* 660 (2019) 1168–81.

29. L. Nitschelm, B. Flipo, J. Auberger, H. Chambaut, S. Dauguet, S. Espagnol, et al. Life cycle assessment data of French organic agricultural products. *Data in Brief*. 38 (2021) 107356.
30. A. Konstantas, H. K. Jeswani, L. Stamford, A. Azapagic. Environmental impacts of chocolate production and consumption in the UK. *Food Research International*. 106 (2018) 1012–25.
31. F. E. Al-Thawadi, Y. W. Weldu, S. G. Al-Ghamdi. Sustainable Urban transportation approaches: Life-cycle assessment perspective of passenger transport modes in qatar. *Transportation Research Procedia*. 48 (2020) 2056–62.
32. B. Cox, W. Jemiolo, C. Mutel. Life cycle assessment of air transportation and the Swiss commercial air transport fleet. *Transportation Research Part D: Transport and Environment*. 58 (2018) 1–13.
33. A. Gallego-Schmid, H. K. Jeswani, J. M. F. Mendoza, A. Azapagic. Life cycle environmental evaluation of kettles: Recommendations for the development of eco-design regulations in the European Union. *Science of the Total Environment*. 625 (2018) 135–46.
34. M. I. Almeida, A. C. Dias, M. Demertzi, L. Arroja. Environmental profile of ceramic tiles and their potential for improvement. *Journal of Cleaner Production*. 131 (2016) 583–93.
35. Latunussa, C. E., F. Ardente, G. A. Blengini, L. Mancini. Life Cycle Assessment of an innovative recycling process for crystalline silicon photovoltaic panels. *Energy Materials and Solar Cells*. 156 (2016) 101–11.
36. M. K. Jaunich, J. DeCarolis, R. Handfield, E. Kemahlioglu-Ziya, S. R. Ranjithan, H. Moheb-Alizadeh. Life-cycle modeling framework for electronic waste recovery and recycling processes. *Resources, Conservation and Recycling*. 161 (2020) 104841.
37. P. Karacal, N. Elginoz, F. Germirli Babuna. Environmental burdens of cataphoresis process. *Desalination and Water Treatment*. 172 (2019) 301–8.
38. H. Kouchaki-Penchah, M. Sharifi, H. Mousazadeh, H. Zarea-Hosseinabadi, A. Nabavi-Pelesaraei. Gate to gate life cycle assessment of flat pressed particleboard production in Islamic Republic of Iran. *Journal of Cleaner Production*. 112 (2016) 343–50.
39. B. A. Türkmen. Cam Ambalaj Üretiminin Çevresel Sürdürülebilirliğinin Değerlendirilmesi. *Bilecik Şeyh Edebali Üniversitesi Fen Bilimleri Dergisi*. 7 (2020) 1026–37.
40. J. B. Guinée, M. Gorrée, R. Heijungs, G. Huppes, R. Kleijn, A. Koning. *Life Cycle Assessment: An Operational Guide to the ISO Standards. Ministry of Housing, Spatial Planning and Environment (VROM) and Centre of Environmental Science (CML)*. Kluwer Academic Publishers, Den Haag and Leiden, The Netherlands, 2001. pp. 2–120.
41. M. M. Asiimwe, I. H. De Kock. An analysis of the extent to which Industry 4.0 has been considered in sustainability or socio-technical transitions. *South African Journal of Industrial Engineering*. 30 (2019) 41–51.
42. J. Vrchota, M. Pech, L. Rolínek, J. Bednář. Sustainability outcomes of green processes in relation to industry 4.0 in manufacturing: systematic review. *Sustainability*. 12 (2020) 5968.
43. L. Rolinek, M. Plevny, J. Kubecova, D. Kopta, M. Rost, J. Vrchota, et al. Level of process management implementation in SMEs and some related implications. *Transactions on Business and Economics*. 14 (2015) 360–77.

44. M. C. Jena, S. K. Mishra, H. S. Moharana. Application of Industry 4.0 to enhance sustainable manufacturing. *Environmental Progress & Sustainable Energy.* 39 (2020) 13360.
45. S. Dzik. COVID-19 convalescent plasma: now is the time for better science. *Transfusion Medicine Reviews.* 34 (2020) 141.
46. S. M. Abdelbasir, C. T. El-Sheltawy, D. M. Abdo. Green processes for electronic waste recycling: A review. *Journal of Sustainable Metallurgy.* 4 (2018) 295–311.
47. R. Brozzi, D. Forti, E. Rauch, D. T. Matt. The advantages of industry 4.0 applications for sustainability: Results from a sample of manufacturing companies. *Sustainability.* 12 (2020) 3647.
48. D. Couckuyt, A. Van Looy. Green BPM as a business-oriented discipline: a systematic mapping study and research agenda. *Sustainability.* 11 (2019) 4200.
49. M. Ghobakhloo. Industry 4.0, digitization, and opportunities for sustainability. *Journal of Cleaner Production.* 252 (2020) 119869.
50. S. Kamble, A. Gunasekaran, N. C. Dhone. Industry 4.0 and lean manufacturing practices for sustainable organisational performance in Indian manufacturing companies. *International Journal of Production Research.* 58 (2020) 1319–37.
51. E. Kamensky. Общество. Личность. Технологии: социальные парадоксы Индустрии 4.0. *Економічний часопис-XXI.* 164 (2017) 9–13.
52. Ministerstvo průmyslu a obchodu (MPO). Iniciativa Průmysl 4.0 https://www.mpo.cz/cz/rozcestnik/ministerstvo/aplikace-zakona-c-106-1999-sb/informace-zverejnovane-podle-paragrafu-5-odstavec-3-zakona/-iniciativa-prumysl-4-0--230485/;Accessed on September 1, 2021. 2016.
53. J. Vrchota, P. Řehoř. Project management and innovation in the manufacturing industry in Czech Republic. *Procedia Computer Science.* 164 (2019) 457–62.
54. V. Mařík. *Průmysl 4.0: výzva pro Českou Republiku.* Management Press, Praha, 2016.
55. M. Švárová, J. Vrchota. Influence of competitive advantage on formulation business strategy. *Procedia Economics and Finance.* 12 (2014) 687–94.

Chapter 7

Lightweight 3D-printed materials

M. Bhuvanesh Kumar
National Institute of Technology, Tiruchirappalli-India

Kongu Engineering College, Erode, India

K. Anand Babu and P. Sathiya
National Institute of Technology, Tiruchirappalli, India

Milon Selvam Dennison
Kampala International University, Western Campus, Uganda

CONTENTS

7.1 INTRODUCTION

Recent decades have seen the emergence of additive manufacturing (AM) techniques as a cost-effective and on-demand manufacturing process for the manufacture of intricate three-dimensional (3D) objects from preestablished geometric data [1]. AM, often known as 3D printing (3DP), is a method that allows for the layer-by-layer production of various components [2]. Other names for the AM technique are "additive fabrication," "fabbing," or "additive layer manufacturing."

The widespread adoption of 3DP followed the introduction of the first 3D printer to the market in 1986 by Charles Hull [3]. The quick development

DOI: 10.1201/9781003252108-7

and manufacturing of small plastic parts were made possible by Charles Hull, who developed the first stereolithography (SLA) manufacturing method. Stereolithography is a technique that utilizes an ultraviolet (UV) laser to set off polymers within a resin (photo polymerization) to create complex three-dimensional parts [4, 5].

A 3DP specialized venture named "3D Systems" first marketed this SLA method in 1987. Since this groundbreaking discovery, much work has been put into developing machines that can process a vast number of materials like polymers and metals. Currently, some of the machines available in the market include fused deposition modeling (FDM) [6, 7], selective laser sintering (SLS), stereolithography (SLA) [8], selectively laser melting (SLM), and direct ink write (DIW) [9, 10] for polymeric and metallic materials. Additionally, inkjet printers employ light to photopolymerize ink drops into intricate shapes [11]. Several research publications have published exhaustive evaluations on these processing and 3DP methods [6, 7, 12–17].

The increasing acceptance of 3DP systems as an alternative to traditional techniques has been attributed to several factors, including the manufacturing of complex geometry with high precision, the most efficient use of materials, design flexibility, and the ability to create personalized products. A simplified illustration of the difference between subtractive and AM methodologies is shown in Figure 7.1. Materials utilized in 3DP now comprise a broad range of elements, including polymers, metals, ceramics, and concrete. Generally, 3DP of composites uses acrylonitrile butadiene styrene (ABS) and polylactic acid (PLA) polymers.

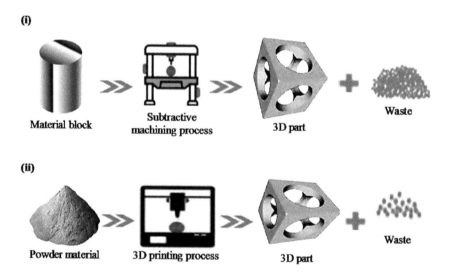

Figure 7.1 Additive vs. subtractive manufacturing processes. (i) 3D object is manufactured by subtracting material from a block of initial material. (ii) 3D printer uses a starting material to construct a 3D component layer by layer.

Traditional methods for processing metals and alloys are longer, more complicated, and costlier than advanced technologies. AM uses concrete as the principal building material, whereas ceramic scaffolds are mostly used in the additive manufacture of structures. More precisely, due to the potential of the AM process, the construction industry has begun making extensive use of 3DP. The WinSun company, for example, successfully mass-printed a set of reasonably inexpensive residence homes in China (USD 4800 per unit) within a single day [18]. The possibilities of large-scale printing, however, are forestalled by inferior mechanical characteristics and anisotropic behavior of 3D-printed objects. As a result, in order to limit the defects and anisotropy of 3D-printed components, it is necessary to choose the most optimal process parameter settings in the 3D printing process. Another factor that might influence the quality of completed products is the printing environment [19].

3DP has caught the attention of many in the medical industry [6]. According to Wohlers Associates, for example, 50% of 3DP in 2020 will be dedicated to creating commercial products [20], including a wide range of medical implants using CT-imaged bio-tissue replicas.

One of AM's key capabilities is the ability to fabricate parts at various sizes from the micro to macro scale. The quality and accuracy of the printed parts, of course, depend on precise measurements, and because of the high resolution, smooth surface, and tightly bonded layers of micro-scale 3DP, further processing steps such as sintering are often necessary [21]. At present, limited amounts of the proper materials available for 3DP prevent wider implementation of the technology, so a wider variety of materials must be developed for use and to enhance the mechanical properties of 3D-printed components.

Hence, this chapter will provide a broad overview of the concepts, principles, parameters, and processes relevant to the design, manufacture, and use of metal and polymer 3D-printed components, as well as their uses in manufacturing.

7.2 NEED AND CONSIDERATIONS OF 3D-PRINTED PARTS

Many procedures, materials, and equipment involved in 3DP have changed over time, which has allowed manufacturers and logistics service providers to use it to revolutionize their respective industries. Currently, AM is used in various sectors that includes construction, prototypes, biomechanics, aerospace, and automobiles [22]. However, despite these advantages (freedom of design, reduced waste, and automation), 3DP adoption in the construction sector has been gradual and confined. As innovative materials and AM processes are progressively being created, new applications are emerging. The primary motivation for this advancement to be accessible to the public is due to earlier patents having expired, allowing manufacturers to invent new 3DP devices.

More precisely, there are two primary considerations when it comes to 3D-printed parts: material and design. In 3DP, five material aspects should be considered: application, aesthetics [23], function, certifications, and cost. With these factors in mind, it's not difficult to keep track of each property required for a part's material and to determine which one best fits the profile. Typically, four types of materials exist that correspond to specific technologies, including photopolymers, powdered thermoplastics, filament thermoplastics, and metals.

Three distinct variables for 3D-printed items must be considered to design successfully for AM: support, topology, and finishing. In today's world, AM is utilized to manufacture fully functional items at a large scale. For example, forward-thinking corporations such as General Electric (GE) and Boeing have adopted AM into their design and manufacturing processes.

7.3 DESIGN PARAMETERS

Recent improvements have lowered the cost of 3D printers, allowing them to be used in schools, libraries, and research labs more widely. Because of its speedy and cost-efficient prototyping potential, 3DP was first primarily employed by designers for creating artistic and functional prototypes. As a result, additional expenses in product development have been minimized using 3DP. Previously, 3DP was used almost exclusively for prototypes and products, but recently the technology has been fully incorporated into many industries, from prototypes to finished products.

Manufacturers have found it difficult to offer customized products, since developing personalized products for end consumers incur high expenses. Alternatively, AM can 3D-print personalized items in small quantities with lower costs, as well as being able to create load-bearing and impact-resistant structures. This is also beneficial in the areas of civil, aircraft, automobile, sport, packaging, and especially the biomedical industry, where specific product requirements must be met [24, 25].

A variety of aspects influences the level of precision that can be achieved in different ways. Firstly, there are the fundamental process aspects such as scaling factor and saturation value. The system maker recommends different values for different materials and purposes for these factors, which should be evaluated for every new build. Several variables affect the overall accuracy more significantly [26]:

- Material selection
- Orientation of the structure
- Geometric aspects and structure, such as open or closed shapes
- Shell and solid wall thicknesses
- Post-treatment measures
- Agent of infiltration

A few essential design considerations should be kept in mind when considering AM. To achieve an ideal balance in the development of any intended item, apart from the above accuracy factors, focus should be given to five elements—sizing, resolution, thickness, orientation, and material choice—which all make your product functional and visually appealing [27].

7.3.1 Size

When designing a product, one factor to consider is scale. Some technologies have a preferred format; some are larger while others are smaller. The Fused Deposition Modeling (FDM) machine can print objects with dimensions as large as 16″ × 14″ × 16″. Any larger item can be sectioned and then bonded together before construction. It is critical to pay attention to the size of individual parts, which directly influences production time. Because building large sections might take longer and use more materials, these expenses will be greater.

7.3.2 Resolution

Resolution is the thickness of each layer of material added during AM. A thinner layer may eliminate flaws or ridges visible on a part's surface, depending on the type of technology utilized. A higher resolution improves the surface quality and the post-processing smoothness. Details are more defined as layers are thinner, while production time increases when layers are thinned out.

SLA (Stereolithography) is for creating high-resolution objects (0.025mm) with a layer thickness up to 25 microns. SLS provides an extremely high level of resolution, around 100 microns (0.1mm). FDM is capable of providing high-resolution with a layer thickness range of 175–250 microns (0.175–0.25mm). Printing in a lesser resolution (330 microns) is feasible. This is a viable alternative for larger, less aesthetic objects that can be created rapidly and inexpensively.

7.3.3 Wall thickness

For an AM component, the wall thickness of the component is an essential design parameter that ensures stability and precision. A thinner portion, however, can be fragile with less precise properties. Thin parts made using SLS are prone to warping, and during the manufacturing process, SLS pieces are exposed to high temperatures and the powder weight. Additionally, materials used in SLS might contract, solidify, and harden as they cool down. The geometric stability of these features is possible when the material has a thickness between 1 to 3 mm. A minimum thickness of 0.4 mm is still achievable. However, achieving geometric stability in additive manufacturing components remains challenging. For instance, a minimum recommended thickness for FDM is 1.6mm.

7.3.4 Orientation

The position of a part on a 3DP platform determines its orientation in 3DP. It can be set vertically, horizontally, or at an angle. A key step in the fabrication process is orienting the components correctly to ensure quality and accuracy, especially as regards geometric dimension and error tolerance. This also has an impact on the amount of energy and resources required. All these contributing factors would influence the process time and cost of the 3DP component.

Additionally, orientation will vary according to the AM technology. Due to the way the layers are produced in FDM, the printed part can withstand higher forces on the X-Y axis but lesser forces on the Z-axis. The drawback is the visibility of lines between layers. In general, for curvy shapes, certain orientations are preferred. Sometimes, the orientation's effectiveness can be enhanced if they protrude from the surface, which necessitates additional support. Unless otherwise stated, designers will position the parts to maximize surface quality and part strength.

7.3.5 Choice of material and design balance

For 3D-printed objects, many materials are available, which are chosen based on the processing method and based on the constraints that apply to the component. Because of this, it is crucial to use an acceptable material when fabricating parts with specific resistances. For example, if a 3D-printed object is subjected to extremely high temperatures or chemical solvents, FDM technology is a good alternative. If a substance with rubber-like qualities is required, flexible resin is the best choice. In contrast, ABS, nylon, and acrylic resin are the best materials for prototyping goods that are reasonably priced. Many variables must be considered in planning a 3D-printed object—including resolution, size, orientation, thickness, and material—all of which will give you a working part that satisfies your requirements.

7.4 MANUFACTURING PROCESS FOR METALS AND POLYMERS

Several AM technologies have been created during the last quarter-century to fulfill the increased demand for printing intricate structures using metals and polymer materials at finer resolutions. Many advancements have contributed to the rise of AM technology, including the capacity to rapidly prototype huge structures that minimize printing flaws and improve mechanical qualities. To get a desired shape and property product using various fabrication processes in AM technology, the following steps are required to produce the 3D object.

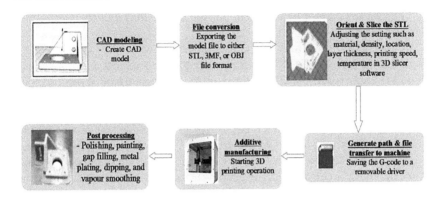

Figure 7.2 Manufacturing stages of 3DP technology.

7.4.1 Steps in the AM process

The 3DP process begins with the division of the intended object's 3D model into layers in a 3D slicer program, taking into account the desired surface accuracy and the generation of the associated G-codes. The data collected during this stage is digitally transferred to a 3D printer, which generates the object's first layer. Subsequent layers are produced on top of the preceding one until all layers have been deposited and the object has been built. Common steps of the 3DP process are shown in Figure 7.2:

- Obtaining the target object's 3D model digitally
- Converting the 3D model into a 3D printer readable digital file format, like STL, and the sliced data in the form of G-codes
- Transmitting the digital information to the 3D printer
- Fabricating the part and performing post-processing treatments if required

a) Modeling of 3D object

To begin the process of 3DP, a 3D model of the object must be created. The model is created utilizing a variety of computer-aided design (CAD) applications and 3D scanning equipment (CT, optical, MRI, or image-based). A design that considers the 3D printer's technology and sensitivity will yield better printing outcomes. The 3D printer's capabilities should also be compatible with the assembly parts and the clearances between moving elements that are to be printed.

b) Conversion of the 3D model file

After acquiring the 3D model, it should be converted into a required file format that can be recognized by the slicer program. In 3DP, the STL file format is the most often used file type to save data. In addition to these formats, many software programs recognize other formats like .obj and .3MF. Once the file format has been converted, the model geometry cannot be changed, but the dimensions and the orientation of the created model can be altered.

c) Orient and slice the STL using 3D slicer software

Before performing 3DP operation, models stored in a suitable format are imported into a 3D slicer program. The main usage of this program is to cut the 3D model into multiple layers and create G-codes that will be supplied into the 3D printer. In addition to the object's position on the 3D printer's table, the program allows you to change the layer thickness, material type density, temperature, and printing speed. Additionally, the slicer software can be used to simulate and monitor the model's construction, exactly as it would in computer-aided manufacturing (CAM) program. This method helps to eliminate errors that could arise during 3DP. Some free 3D printer slicing tools, such as the open-source software UltimakerCura (also known as Ultimaker B.V.) and CraftWare (also known as CraftUnique), are accessible for everyone to use. The generated codes can be sent to the printer through a wired/wireless network or a memory card.

d) Process of printing 3D object

As soon as the file transfer is complete, the 3D printer recognizes the codes and begins depositing once it reaches the proper temperature level. Before starting the printing process, the printer should be leveled, calibration operations conducted, and the material correctly prepared for printing. The duration of the print varies based on the 3DP technology, the material density, size and geometry of the model, the quantity of support required, and the preferred resolution level.

In AM, the object's geometry directly affects the printing time. The objects with complex shapes are critical to position in the printing table. The time required to isolate the model from support structures can be reduced when the objects are positioned in such a way that the least amount of support structure is necessary.

e) Post-processing treatment

After printing, the part is removed from the tray and the support structures are wiped away by cleaning. There are two basic types of support structures: conventional and dissolvable. Using proper hand tools, conventional support structures can be cleaned from the workpiece directly, and dissolvable structures can be removed by submerging the workpiece in water or using material-specific solutions. After cleaning, the post-processing activities such as sanding, epoxy coating, polishing, gap filling, painting, metal plating, dipping, and vapor smoothing may be performed if required.

7.4.2 Types of the AM process

In general, fused deposition modeling (FDM) is the most prevalent type of 3DP that usually uses polymer filaments. In addition to FDM, other AM methods such as SLS, SLM, and liquid binding with 3DP, stereolithography, inkjet printing, DED, LOM, and contour crafting for processing different materials [28–30] are used.

a) FDM process

A key development in the field of AM was introduced in the early 1990s, when Scott Crump, a co-founder of Stratasys, patented the FDM process. FDM is an AM process that is classified within the family of material extrusion generally known as Fused Filament Fabrication (FFF). In this instance, a thin layer of thermoplastic wire filament is employed to 3D-print material layers as shown in Figure 7.3. The materials utilized in this method are thermoplastic polymers in the form of filaments such as ABS, PLA, Nylon, PETG, TPU, and PEI.

To achieve a semi-liquid condition, the filament is heated at the nozzle to get it up to temperature. Once it's ready, it is extruded onto the platform or on top of previously printed layers. This method necessitates the thermo-plasticity of the polymer filament to fuse during printing, solidify at room temperature after printing, and remain connected until the end of the print. To control the final mechanical qualities of printed objects, the layer thickness, width, and orientation of the filaments and air gap are key process parameters [31]. A fundamental factor contributing to mechanical weakness is inter-layer distortion [32].

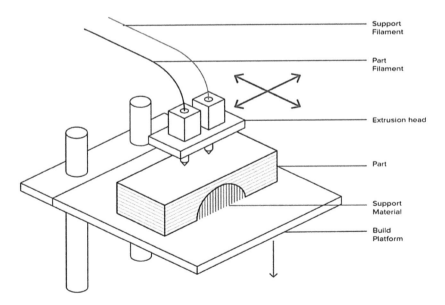

Support Filament

Part Filament

Extrusion head

Part

Support Material

Build Platform

Figure 7.3 Schematic layout of a typical FDM method setup [35]. https://facfox.com/docs/kb/introduction-to-fdm-3d-printing.

One of the key benefits of FDM is its low cost, high printing rates, and ease of the process. Alternatively, the main limitations of FDM are its lack of strong mechanical qualities, low surface quality [33], and limited thermoplastic materials [31]. The development of FDM-based fiber-reinforced composites has improved the mechanical properties of 3D-printed parts. However, the orientation of the fibers, the bonding of the fibers to the matrix, and the creation of voids are the primary issues encountered in 3D-printed composite parts [4, 34].

b) SLS, SLM, and Binder Jetting processes

The SLS and SLM techniques are part of the Powder Bed Fusion (PBF) family of AM techniques. The process of selective laser sintering (SLS) involves fusing polymer powder particles in layers, which results in the construction of a part as shown in Figure 7.4a. An ultraviolet laser beam or a binder is employed to fuse the particles in each layer. The final 3D part is constructed by rolling successive layers of powder on top of each other and fusing them. A vacuum is used to remove the extra powder, and if necessary, additional processing and details such as coating, sintering, or infiltration are carried out on the piece of metal.

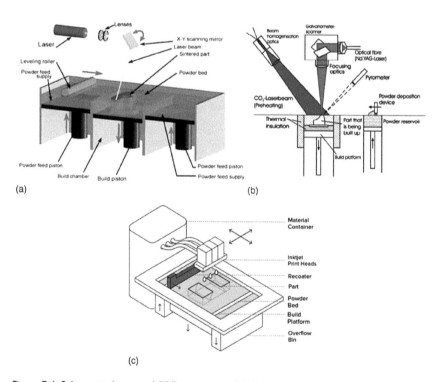

Figure 7.4 Schematic layout of PBF processes, (a) Selective laser sintering process [39], (b) Selective laser melting process [40], (c) Binder jetting process [41]. https://www.hubs.com/knowledge-base/how-design-parts-binder-jetting-3d-printing/.

The efficacy of this method depends primarily on the powder density and packing that determine the density of the printed part [36]. Only powders possessing low melting/sintering temperature can be used with the laser; otherwise, a liquid binder should be used. Many types of polymers, metals, and alloy powders (in granular form) can be used in the SLS process.

While SLM is effective on certain metals like steel and aluminum, it is not feasible for use with other metals. SLS laser scanning does not melt the powders to the point where they fuse into the molecular matter, and the elevated temperature on the surface of the grains causes the grains to fuse at the molecular level. In contrast, particles after laser scanning in SLM as shown in Figure 7.4b are entirely melted and completely intermixed, resulting in improved mechanical characteristics [37]. SLM can be found in reference [38] for a thorough exploration of diverse materials and application scenarios.

The technology is referred to as 3DP when using a liquid binder for fabricating a 3D object as shown in Figure 7.4c. 3DP is strongly dependent on many other variables, such as the kind of binder, the type of powder particles, the deposition rate, and the interactions among the powder and binder [4, 36]. Binder deposition tends to produce parts with higher porosity than other methods like SLS or SLM, which can produce dense parts [36]. One of the major influences on the sintering process is the power and speed of laser scanning. According to Lee et al. [37], various types of lasers are available, which have different influences on the 3DP process.

In general, PBF has excellent resolution and quality printing. It's therefore suited for printing intricate structures. This approach is utilized in advanced applications, like tissue engineering scaffolds, lattices, electronics, and aerospace industries. This technique has the advantage of utilizing the powder bed as a support, overcoming challenges encountered in removing supporting material. However, PBF, which is a lengthy process, has two significant drawbacks: it is expensive, and the powder bed has high porosity when it is fused with a binder.

c) The inkjet printing and contour crafting processes

The inkjet printing process is one of the important methods for the AM of ceramics, which belongs to the material jetting family. Scaffolds for tissue engineering application requires advanced and intricate ceramic structures for which inkjet printing is a perfect alternative.

In inkjet printing, the ceramic powders suspended in a suspension medium (e.g., ZrO_2 suspended in water) [42] are forced by an injection nozzle as droplets and deposited over the substrate in a required pattern to form a layer. When the droplets solidify, they form a solid layer and provide sufficient strength to support and hold the layers to be printed in subsequent passes, as shown in Figure 7.5. Compared to

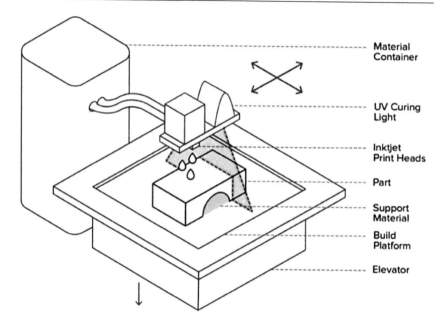

Figure 7.5 Inkjet printing process [45]. https://www.hubs.com/knowledge-base/how-design-parts-material-jetting-3d-printing/.

other AM techniques, the inkjet printing method is quicker and offers better flexibility to design and fabricate complex structures efficiently.

Two types of ceramic inks are available, wax- and liquid-based suspensions. Both have different principles for solidification. The wax-based inks solidify based on the substrate's coldness. Liquid suspension solidifies based on the evaporation principle.

The quality of inkjet-printed parts depends on certain parameters, such as size and distribution of particles, solid content percentage, liquid viscosity, and the rate of extrusion. The nozzle size determines the extrusion rate and printing speed [43]. On the other hand, few drawbacks such as lack of adhesion, resolution, and maintaining workability have been reported by previous studies.

Similar to inkjet printing technology, contour crafting, as shown in Figure 7.6, deposits layers of materials based on droplets extruded from a nozzle, the difference being its capability to build large structures. Contour crafting can extrude and deposit concrete or soil through high pressure nozzles to build structures. This technology has been used effectively in civil engineering for construction [44].

d) SLA process

The SLA method was invented and patented in early 1986. It is well-known for being the first 3DP method, and it is an important AM process that is part of the Vat Photopolymerization [47] family. With the

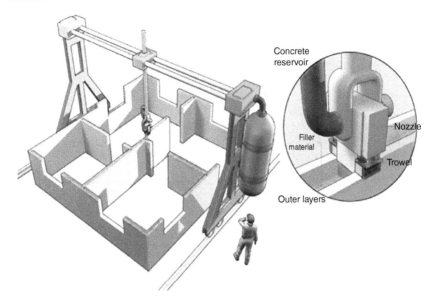

Figure 7.6 Contour crafting process [46].

SLA technique, an object is made by applying layers of polymer resin to a UV laser-cured resin. As soon as the resin or monomer solution is exposed to UV light, a chain reaction is set in motion. Most monomers are UV-responsive and can rapidly form polymer chains once they have been activated, which is known as radicalization. The design inside the resin layer is hardened during polymerization to ensure that the subsequent layers adhere properly, as shown in Figure 7.7. The unreacted resin is scraped off when the printing process is completed.

To obtain the necessary mechanical performance, a post-processing treatment such as photo-curing or heating may be employed for some printed parts. The polymer materials used in SLA are photosensitive and are a thermosetting type that come in a liquid form. If particularly accurate items or a smooth surface finish is required, SLA can be the most cost-effective 3DP technique available.

e) The Direct Energy Deposition (DED) process

The DED approach [13] has been utilized to manufacture 3D parts of high-performance super-alloys. Additionally, this approach is referred to by multiple names according to their different working principles, such as laser engineered net shaping (LENS) [48], directed light fabrication (DLF), direct metal deposition (DMD), laser solid forming (LSF), electron beam additive manufacturing (EBAM) [49], and wire and arc additive manufacturing (WAAM).

DED, as shown in Figure 7.8, is a method of melting a feedstock material (wire or powder) by using an electron or laser beam energy

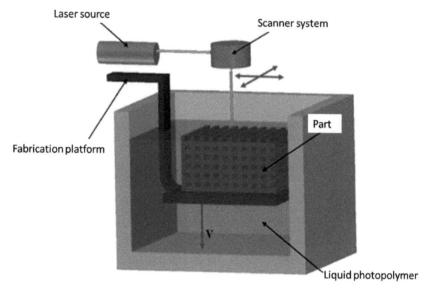

Figure 7.7 Schematic diagram of Stereolithography process [39].

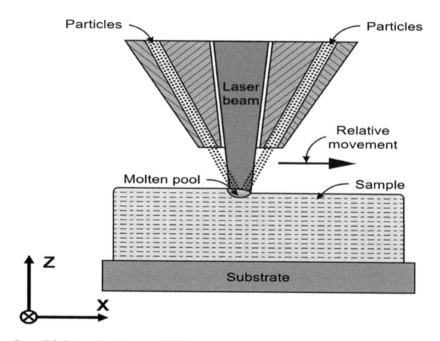

Figure 7.8 Schematic diagram of DED method [50].

source that is focused directly on a small portion of the substrate and melts a small area in the substrate. After the movement of the laser beam, the melted material is deposited and fused into the melted substrate, which is subsequently solidified [13]. DED is like FDM, except instead of depositing the material layer-by-layer, the feedstock is melted before laying down the material. As a result, it may be useful for filling cracks and retrofitting produced pieces for which the powder-bed approach is ineffective.

This approach is ideal for applying diverse materials on multiple axes simultaneously. More precisely, conventional subtractive methods can also be used with DED to complete machining. This process is frequently utilized in aerospace applications using titanium, Inconel, stainless steel, and aluminum alloys. However, when compared to SLS or SLM, it has less precision, lower surface quality, and can only make objects that are less complex [13]. As a result, DED is often utilized for large components with little complexity, as well as for repairing larger components of a similar size.

DED has the potential to minimize manufacturing time and costs while providing good mechanical characteristics, controlled microstructure, and precise composition control, among other benefits. This process is suitable for specific applications in numerous industries, including the automotive and aerospace industries.

f) Laminated object manufacturing (LOM)

The commercial availability of LOM was an important milestone in the progression of AM because it (Figure 7.9) was the first practical process based on layer-by-layer cutting and lamination of sheets or rolls of materials. To create successive layers, a mechanical cutter or laser is used to cut through the layers, and they have then adhered together or vice versa (bond-then-form). Using the form-then-bond process, the bonding of ceramics and metallic materials is greatly simplified because superfluous materials are removed first. After cutting, the surplus materials are left for support, and they can be removed and recycled once the operation is complete [13]. Many materials can be incorporated into LOM such as ceramics, paper, polymers, and metal-filled tapes.

Based on the type of materials and desired qualities, post-processing procedures such as high-temperature treatment can be utilized following fabrication. LOM has seen extensive use in many industries such as paper manufacture, electronics, foundry industries, and structures that use both complex and simple forms of programming. Structures with sensors and processors are known as smart structures. LOM can reduce the cost and manufacturing time required for large-scale products and is an effective AM process for these products. However, without post-processing, the LOM has poor surface quality and lower dimensional accuracy compared to PBF methods. Additionally,

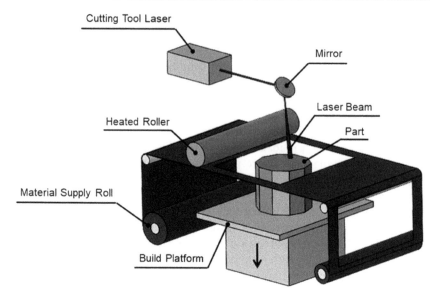

Figure 7.9 Schematic diagram of LOM method [51].

removing surplus laminates after the object is formed is more time-intensive than with the PBF process. Therefore, it is not advised for complex geometries.

7.5 APPLICATIONS

Innovations in AM processes have improved significantly in recent years, resulting in a wider range of industrial applications. AM is particularly well suited to generating low volumes of components, particularly for parts with complex geometry, when compared to subtractive manufacturing. AM lends itself to customization, such as making personalized hip and knee replacement inserts. AM applications can be found in the fields below:

a) Aerospace industry
The aircraft industry is a strong potential prospect for AM. In aerospace applications, lightweight high strength materials are generally required. Because of the significant reduction in design cost, higher flexibility of design, and a greater diversity of goods for customers, AM is extremely important. It can be used to design and fabricate complex, strong, lightweight structures with lower costs and shorter lead times.

In place of conventional molding and machining, the aerospace industry has shifted to 3DP for small-scale production. This is expected to impact the size and weight of aerospace machines by replacing small components with 3D-printed parts, such as arm rests, food trays, and seat belts, to name just a few. As a result, 3DP may be utilized to build hierarchical structures that are strong and efficient in terms of material utilization [52, 53].

Furthermore, AM is not only useful for repairing worn-out parts, but also for building replacements when needed. The capacity to produce on-demand spare components decreases production costs and saves warehousing space [54]. In recent years, NASA has developed a rover, known as Desert RATS, that can support humans in utilizing a pressurized chamber [55]. The rover will ferry human beings to Mars—it is a 3D-printed spacecraft that includes 70 individual parts that consist of flame-retardant vents and housings, big pod doors, camera mounts, complicated electronics, and others. Materials such as ABS, PCABS, and PC were used to 3D print the rover's components [55].

Many firms have moved to AM, particularly those that seek to improve their speed of manufacturing without changing their commodities [52]. This change is due to the volatile market and low cost of producing such small objects. Many hurdles must be solved in order to promote AM's progress; for example, the current AM machine speed for bulk production is slow, fewer polymeric material alternatives are available, and current machines prevent the fabrication of large products [47, 52]. Modern AM techniques can be utilized to modify the mechanical, physical, wear, and thermal response of functionally graded materials, metal matrix, and polymer-based composites and superalloys. In addition, on-demand manufacturing reduces costs and minimizes potential storage damage [30].

b) Electronic devices industry

To perform functions such as reduced energy consumption, sophisticated miniaturization, and added smart capabilities, electronic devices require the proper design and materials for manufacture and components that possess mechanical, geometrical, and optical functions [56]. Because of the continuous evolution of technology, manufacturing prototypes and finished products must adapt rapidly. Electronic devices are typically manufactured by using subtractive procedures in which sacrificial elements are masked and etched away [57], but AM reduces energy use, material waste, and processing steps and time.

Three-dimensional printing is being utilized for stand-in processes in the mounting and assembly of electronic devices [58]. The demand for smaller parts is fulfilled because of AM; for example, thin films [59], inductors [60], solar cells [61], and other applications. The most frequent 3DP processes for the electronics industry are inkjet and direct inscription of conductive inks.

According to Jennifer Lewis et al. (2015), using a 3DP procedure, they created a quantum dot LED system with embedded green and orange-red light emitters in a silicone matrix [62]. Similarly, Lee Y et al. (2008) developed an ink for inkjet printing employing a copper nanoparticle stabilized with polyvinyl pyrrolidine and 2-(2-butoxye-thoxy) ethanol [63]. The ink was sintered at 200°C after being printed on a polyimide substrate. The resulting electronic device exhibited low electrical resistance.

Bionic ears are also printable using an inkjet printer [64]. The inks are made up of alginate-chondrocyte hydrogel matrices, conductive silicone-silver nanoparticle polymers, and cell-cultured alginate. The 3D-printed ears have improved auditory sensing for radio-frequency reception, making it possible to listen to stereo music. This outcome shows that electronics and bioengineering can be combined to achieve high-tech breakthroughs.

Future electronics R&D will make use of inexpensive methods, the ability to flexibly design items, and the speed of 3D printers for creating prototypes. For example, producing circuit boards with the printer will afford greater accuracy and flexibility while potentially saving money, improving the environment, and expediting the production process. Adaptive 3DP, which combines real-time feedback control with DIW of functional materials to manufacture devices on dynamic surfaces, is also an intriguing study subject [65]. This 3DP approach could pave the way for new smart manufacturing technologies for directly manufactured wearable devices and new opportunities in biological and biomedical research, as well as the study and treatment of advanced medical treatments.

c) Biomedical applications
d) In dental and implant medicine, AM is widely believed to have an impressive market, notably in terms of the fabrication of patient-specific complicated components, high material usage, tissue growth support, and personalized service for individual patients. With materials like metals, polymers, ceramics, and bio-inks, this technology has the potential to be adopted in research and clinical settings for both personal and medical applications [1, 66]. The application of 3DP technology for use in regenerative medicine to produce tissue constructs created with cell-laden bio-inks is known as bioprinting. Another advanced technology in tissue engineering and regenerative medicine is 3D bioprinting, which can create complex tissues to resemble native organs and tissues [67]. Figure 7.10 shows some biomedical products produced using 3DP technology.

"Fusion Deposition Modeling (FDM)" was described by Malafaya BA, et al. [64] as a technique with a growing number of applications in biomedical engineering such as developing customized and functional medical devices. Due to the growing trend over the last decade to fabricate bones, teeth, and cartilage, as well as softer tissues like organs and skin, 3DP has seen increased usage [2–4, 16, 37].

Figure 7.10 Biomedical products by using 3DP technology [68].

In regenerative medicine, a porous scaffold is inserted into the patient to serve as a template for tissue regeneration as the implant slowly degrades in the body (other implants remain in place for the duration of the patient's life).

3DP enables the rapid production of bespoke prosthetics and implants with precise architectural designs. In order to create the structure, x-ray, MRI, and CT images must be translated into STL file formats. The STL file can be processed by software to create a design tailored to the patient's demands. Bone regeneration prostheses are often made from metallic materials; ABS and PLA are the best polymers for scaffold production since they are non-biodegradable.

The dental industry employs 3DP technologies for a variety of applications, including restorative dentistry, dental implants, and orthodontics. At present, dental experts have access to 3D printers, and it is possible to produce designs in a clinical setting. By using CT scans, dentists may swiftly create and replace lost teeth by creating a replica of the patient's anatomy. 3DP is utilized for the fabrication of aligners, braces, dental implants, and crowns.

7.6 CONCLUSION

Advancements in the field of manufacturing have led to the invention of AM technology. The refinement of the process parameters has helped achieve better mechanical properties in lightweight materials. Through AM we have realized a reduction in cost and processing time, manufactured complex shapes that have never been produced with the traditional subtractive fabrication process, and reduced geometrical and dimensioning errors. Diverse materials such as titanium alloys, nickel-based alloys, stainless-steel, and alumina have all been used in various processes to produce products critical to biological components without a compromise in quality function and performance. However, there are certain limitations to this process that include factor-like cost, layer misalignment, anisotropic mechanical property, development of residual stress, etc. With the invention of new hybrid manufacturing techniques, this appears to be the future for AM technologies. Cutting-edge research conducted across leading research labs also predict that the future of AM lies in hybridization. One such futuristic technique will be the implementation of Computer-Aided Process Planning (CAPP), along with non-destructive testing systems, which can detect defects during the production stage itself. The potential implementation of these advanced techniques will increase the visibility and application of AM manufactured parts to a greater extent in the near future.

REFERENCES

1. P. Ahangar, M. E. Cooke, M. H. Weber, and D. H. Rosenzweig, Current Biomedical Applications of 3D Printing and Additive Manufacturing, *Appl. Sci.* 9 (2019), p. 1713.
2. G. W. Bishop, *3D Printed Microfluidic Devices*, in *Microfluidics for Biologists*, Springer International Publishing, Cham, 2016, pp. 103–113.
3. C. W. Hull, *Apparatus for Production of Three-Dmensonal Objects By Stereo Thography*, Pat. US4575330A 19 (1986), pp. 1–4.
4. X. Wang, M. Jiang, Z. Zhou, J. Gou, and D. Hui, 3D printing of polymer matrix composites: A review and prospective, *Compos. Part B Eng.* 110 (2017), pp. 442–458.
5. S. C. Ligon, R. Liska, J. Stampfl, M. Gurr, and R. Mülhaupt, Polymers for 3D Printing and Customized Additive Manufacturing, *Chem. Rev.* 117 (2017), pp. 10212–10290.
6. J. W. Stansbury and M. J. Idacavage, 3D printing with polymers: Challenges among expanding options and opportunities, *Dent. Mater.* 32 (2016), pp. 54–64.
7. S. Ford and M. Despeisse, Additive manufacturing and sustainability: an exploratory study of the advantages and challenges, *J. Clean. Prod.* 137 (2016), pp. 1573–1587.
8. K. V. Wong and A. Hernandez, A review of additive manufacturing, *ISRN Mech. Eng.* 2012 (2012), pp. 1–10.

9. J. A. Lewis, Direct ink writing of 3D functional materials, *Adv. Funct. Mater.* 16 (2006), pp. 2193–2204.

10. E. B. Duoss, T. H. Weisgraber, K. Hearon, C. Zhu, W. Small, T. R. Metz, et al., Three-dimensional printing of elastomeric, cellular architectures with negative stiffness, *Adv. Funct. Mater.* 24 (2014), pp. 4905–4913.

11. S. F. S. Shirazi, S. Gharehkhani, M. Mehrali, H. Yarmand, H. S. C. Metselaar, N. Adib Kadri, et al., A review on powder-based additive manufacturing for tissue engineering: Selective laser sintering and inkjet 3D printing, *Sci. Technol. Adv. Mater.* 16 (2015), p. 033502.

12. B.-J. de Gans, P. C. Duineveld, and U. S. Schubert, Inkjet printing of polymers: State of the art and future developments, *Adv. Mater.* 16 (2004), pp. 203–213.

13. I. Gibson, D. Rosen, and B. Stucker, *3D Printing, Rapid Prototyping, and Direct Digital Manufacturing*, 2nd edition, Springer, New York, NY, 2015.

14. B. C. Gross, J. L. Erkal, S. Y. Lockwood, C. Chen, and D. M. Spence, Evaluation of 3D printing and its potential impact on biotechnology and the chemical sciences, *Anal. Chem.* 86 (2014), pp. 3240–3253.

15. B. N. Turner, R. Strong, and S. A. Gold, A review of melt extrusion additive manufacturing processes: I. Process design and modeling, *Rapid Prototyp. J.* 20 (2014), pp. 192–204.

16. T. Jungst, W. Smolan, K. Schacht, T. Scheibel, and J. Groll, Strategies and molecular design criteria for 3D printable hydrogels, *Chem. Rev.* 116 (2016), pp. 1496–1539.

17. I. Gibson, D. W. Rosen, and B. Stucker, *Additive Manufacturing Technologies*, Springer, US, Boston, MA, 2010.

18. P. Wu, J. Wang, and X. Wang, A critical review of the use of 3-D printing in the construction industry, *Autom. Constr.* 68 (2016), pp. 21–31.

19. O. Ivanova, C. Williams, and T. Campbell, Additive manufacturing (AM) and nanotechnology: Promises and challenges, *Rapid Prototyp. J.* 19 (2013), pp. 353–364.

20. B. Berman, 3D printing: The new industrial revolution, *Bus. Horiz.* 55 (2012), pp. 155–162.

21. M. Vaezi, H. Seitz, and S. Yang, A review on 3D micro-additive manufacturing technologies, *Int. J. Adv. Manuf. Technol.* 67 (2013), pp. 1721–1754.

22. S. Varatharaj Kannan, S. T. Yogasundar, R. Suraj Singh, S. Tamilarasu, and M. Bhuvanesh Kumar, Rapid prototyping of human implants with case study, *Int. J. Inn. Res. Sci. Eng. Tech* 3 (2014), pp. 319–324.

23. M. Galati, P. Minetola, G. Marchiandi, E. Atzeni, F. Calignano, A. Salmi, et al., A methodology for evaluating the aesthetic quality of 3D printed parts, *Procedia CIRP* 79 (2019), pp. 95–100.

24. S. M. Sajadi, C. F. Woellner, P. Ramesh, S. L. Eichmann, Q. Sun, P. J. Boul, et al., 3D printed tubulanes as lightweight hypervelocity impact resistant structures, *Small* 15 (2019), pp. 1904747.

25. S. Yuan, C. K. Chua, and K. Zhou, 3D-printed mechanical metamaterials with high energy absorption, *Adv. Mater. Technol.* 4 (2019), pp. 1–9.

26. D. Dimitrov, W. van Wijck, K. Schreve, and N. de Beer, Investigating the achievable accuracy of three dimensional printing, *Rapid Prototyp. J.* 12 (2006), pp. 42–52.

27. Fablab Inc. *5 important elements to consider when printing in 3D*, In: Home » Blogs.https://www.fablabinc.com/en/blog/5-important-elements-consider-when-printing-3d

28. B. Bhushan and M. Caspers, An overview of additive manufacturing (3D printing) for microfabrication, *Microsyst. Technol.* 23 (2017), pp. 1117–1124.

29. D. Ortiz-Acosta and T. Moore, Functional 3D printed polymeric materials, in *Functional Materials*, Dipti Sahu (Ed.), IntechOpen, 2018, London, United Kingdom. [Online]. Available: https://www.intechopen.com/chapters/63193 doi: 10.5772/intechopen.80686

30. T. D. Ngo, A. Kashani, G. Imbalzano, K. T. Q. Nguyen, and D. Hui, Additive manufacturing (3D printing): A review of materials, methods, applications and challenges, *Compos. Part B Eng.* 143 (2018), pp. 172–196.

31. O. A. Mohamed, S. H. Masood, and J. L. Bhowmik, Optimization of fused deposition modeling process parameters: A review of current research and future prospects, *Adv. Manuf.* 3 (2015), pp. 42–53.

32. A. K. Sood, R. K. Ohdar, and S. S. Mahapatra, Parametric appraisal of mechanical property of fused deposition modelling processed parts, *Mater. Des.* 31 (2010), pp. 287–295.

33. J. S. Chohan, R. Singh, K. S. Boparai, R. Penna, and F. Fraternali, Dimensional accuracy analysis of coupled fused deposition modeling and vapour smoothing operations for biomedical applications, *Compos. Part B Eng.* 117 (2017), pp. 138–149.

34. P. Parandoush and D. Lin, A review on additive manufacturing of polymer-fiber composites, *Compos. Struct.* 182 (2017), pp. 36–53.

35. A. B. Varotsis, Introduction to FDM 3D printing, In: Knowl. Base, Manuf. Process. Explain. 2021. https://www.hubs.com/knowledge-base/introduction-fdm-3d-printing/

36. B. Utela, D. Storti, R. Anderson, and M. Ganter, A review of process development steps for new material systems in three dimensional printing (3DP), *J. Manuf. Process.* 10 (2008), pp. 96–104.

37. H. Lee, C. H. J. Lim, M. J. Low, N. Tham, V. M. Murukeshan, and Y.-J. Kim, Lasers in additive manufacturing: A review, *Int. J. Precis. Eng. Manuf. Technol.* 4 (2017), pp. 307–322.

38. C. Y. Yap, C. K. Chua, Z. L. Dong, Z. H. Liu, D. Q. Zhang, L. E. Loh, et al., Review of selective laser melting: Materials and applications, *Appl. Phys. Rev.* 2 (2015), p. 041101.

39. M. Bhuvanesh Kumar and P. Sathiya, Methods and materials for additive manufacturing: A critical review on advancements and challenges, *Thin-Walled Struct.* 159 (2021), p. 107228.

40. H. Yves-Christian, W. Jan, M. Wilhelm, W. Konrad, and P. Reinhart, Net shaped high performance oxide ceramic parts by selective laser melting, *Phys. Procedia* 5 (2010), pp. 587–594.

41. A. B. Varotsis, Introduction to binder jetting 3D printing, In: Knowl. Base, Manuf. Process. Explain. 2021. https://www.hubs.com/knowledge-base/introduction-binder-jetting-3d-printing/

42. R. Dou, T. Wang, Y. Guo, and B. Derby, Ink-Jet Printing of Zirconia: Coffee Staining and Line Stability, *J. Am. Ceram. Soc.* 94 (2011), pp. 3787–3792.

43. N. Travitzky, A. Bonet, B. Dermeik, T. Fey, I. Filbert-Demut, L. Schlier, et al., Additive manufacturing of ceramic-based materials, *Adv. Eng. Mater.* 16 (2014), pp. 729–754.

44. B. Khoshnevis, Automated construction by contour crafting—related robotics and information technologies, *Autom. Constr.* 13 (2004), pp. 5–19.

45. A. B. Varotsis, Introduction to material jetting 3D printing, In: Knowl. Base, Manuf. Process. Explain. 2021. https://www.hubs.com/knowledge-base/introduction-material-jetting-3d-printing/

46. H. Patil, S. Nalavade, and N. Pisal, Contour crafting (A Management Tool for Swift Construction), *Int. Res. J. Eng. Technol.* 06 (2019), pp. 837–840.

47. W. Gao, Y. Zhang, D. Ramanujan, K. Ramani, Y. Chen, C. B. Williams, et al., The status, challenges, and future of additive manufacturing in engineering, *Comput. Des.* 69 (2015), pp. 65–89.

48. K. S. Prakash, T. Nancharaih, and V. V. S. Rao, Additive manufacturing techniques in manufacturing: An overview, *Mater. Today Proc.* 5 (2018), pp. 3873–3882.

49. D. Herzog, V. Seyda, E. Wycisk, and C. Emmelmann, Additive manufacturing of metals, *Acta Mater.* 117 (2016), pp. 371–392.

50. F. Arias-González, A. Rodríguez-Contreras, M. Punset, J. M. Manero, Ó. Barro, M. Fernández-Arias, et al., In-situ laser directed energy deposition of biomedical Ti-Nb and Ti-Zr-Nb alloys from elemental powders, *Metals (Basel).* 11 (2021), pp. 1205.

51. A. Razavykia, E. Brusa, C. Delprete, and R. Yavari, An overview of additive manufacturing technologies: A review to technical synthesis in numerical study of selective laser melting, *Materials (Basel).* 13 (2020), pp. 3895.

52. J. Coykendall, M. Cotteleer, L. Holdowsky, and M. Mahto, *3D Opportunity in Aerospace and Defense*, Deloitte University Press, UK (2014), pp. 1–28.

53. M. Hambach, M. Rutzen, and D. Volkmer, Properties of 3D-printed fiber-reinforced portland cement paste, in *3D Concrete Printing Technology*, Elsevier, UK, 2019, pp. 73–113.

54. P. Pecho, V. Ažaltovič, B. Kandera, and M. Bugaj, Introduction study of design and layout of UAVs 3D printed wings in relation to optimal lightweight and load distribution, *Transp. Res. Procedia* 40 (2019), pp. 861–868.

55. Stratasys, *3D Printing a Space Vehicle*, Internet. (2013), pp. ISO 9001:2008.

56. H. Ota, S. Emaminejad, Y. Gao, A. Zhao, E. Wu, S. Challa, et al., Application of 3D printing for smart objects with embedded electronic sensors and systems, *Adv. Mater. Technol.* 1 (2016), p. 1600013.

57. H. W. Tan, T. Tran, and C. K. Chua, A review of printed passive electronic components through fully additive manufacturing methods, *Virtual Phys. Prototyp.* 11 (2016), pp. 271–288.

58. D. Espalin, D. W. Muse, E. MacDonald, and R. B. Wicker, 3D printing multifunctionality: Structures with electronics, *Int. J. Adv. Manuf. Technol.* 72 (2014), pp. 963–978.

59. H. Okimoto, T. Takenobu, K. Yanagi, Y. Miyata, H. Shimotani, H. Kataura, et al., Tunable carbon nanotube thin-film transistors produced exclusively via inkjet printing, *Adv. Mater.* 22 (2010), pp. 3981–3986.

60. W. Liang, L. Raymond, and J. Rivas, 3D-printed air-core inductors for high-frequency power converters, *IEEE Trans. Power Electron.* 31 (2016), pp. 52–64.

61. S. H. Eom, S. Senthilarasu, P. Uthirakumar, S. C. Yoon, J. Lim, C. Lee, et al., Polymer solar cells based on inkjet-printed PEDOT:PSS layer, *Org. Electron.* 10 (2009), pp. 536–542.

62. J. A. Lewis and B. Y. Ahn, Three-dimensional printed electronics, *Nature* 518 (2015), pp. 42–43.

63. Y. Lee, J. Choi, K. J. Lee, N. E. Stott, and D. Kim, Large-scale synthesis of copper nanoparticles by chemically controlled reduction for applications of inkjet-printed electronics, *Nanotechnology* 19 (2008), p. 415604.

64. B. A. Malafaya, M. C. Marques, I. A. Ferreira, M. M. F. Machado, G. A. R. Caldas, J. Belinha, et al., Additive Manufacturing from a Biomedical Perspective, in *2019 IEEE 6th Portuguese Meeting on Bioengineering (ENBENG)*, 2019, pp. 1–4.

65. Z. Zhu, S.-Z. Guo, T. Hirdler, C. Eide, X. Fan, J. Tolar, et al., 3D printed functional and biological materials on moving freeform surfaces, *Adv. Mater.* 30 (2018), p. 1707495.

66. Y. Liu, W. Wang, and L.-C. Zhang, Additive manufacturing techniques and their biomedical applications, *Fam. Med. Community Heal.* 5 (2017), pp. 286–298.

67. J. Gopinathan and I. Noh, Recent trends in bioinks for 3D printing, *Biomater. Res.* 22 (2018), p. 11.

68. N. Shahrubudin, P. Koshy, J. Alipal, M. H. A. Kadir, and T. C. Lee, Challenges of 3D printing technology for manufacturing biomedical products: A case study of Malaysian manufacturing firms, *Heliyon* 6 (2020), p. e03734.

Chapter 8

Biodegradable materials

Akesh B. Kakarla, Satya G. Nukala, and Ing Kong

La Trobe University, Bendigo, Australia

CONTENTS

8.1 INTRODUCTION

Rapid industrialization and urbanization continue to evolve and refine the range and requirements of manufactured goods. The refinements must meet the increasing demands of customer requirements, recovery, and environmental credentials. However, the modifications also decrease energy requirements and meet ecological challenges. Thus, the application of lightweight environmentally friendly materials (LWEFMs) in the automobile and aerospace industries has grown exponentially from the previous decade [1–4].

DOI: 10.1201/9781003252108-8

LWEFMs are classified as solid materials characterized by very low density (100 mg/cm^3), high porosity (>50%), large specific surface area, and various pore structures [2]. For example, considering LWEFM products in automobiles helps decrease vehicle weight and fuel consumption [5]. LWEFMs are mostly aluminum alloys, high strength steel, magnesium alloy, polymers, and composites. However, LWEFMs impose a significant challenge to the industry in terms of selecting suitable materials and manufacturing processes and considering property consideration and performance evaluation [4]. Unsuitable materials cause damages and losses to the industry in the long term, making material selection a salient issue. Selected materials have been analyzed regarding their physical, mechanical, and magnetic properties; raw material price; product configuration and environmental effect; and market trends and availability.

Traditional polymer matrices such as polyester, epoxy, and vinyl ester are attractive for lightweight applications, as they provide good mechanical properties compared to biodegradable polymers [4]. Additionally, these polymers are reinforced with glass fibers, carbon fiber, clay, silica, and other non-degradable fillers to produce composite materials [1, 3, 4]. For instance, carbon fiber-reinforced epoxy composites provide high strength-to-weight ratio, high tensile modulus-to-weight ratio, high rigidity, and fatigue strength. Despite certain advantages of synthetic fiber composites, deteriorating environmental conditions are the primary concern in using them [1, 6].

In recent years, research and development have focused on developing more environmentally friendly, cost-effective, and sustainable materials. In this context, the use of natural fiber-reinforced polymer composites (NFRPCs) as alternative materials to synthetic fiber composites has increased because of their potential ecological attributes and mechanical properties [1]. NFRPCs have many advantages: biodegradable, compostable, eco-friendly, non-toxic, lightweight, rigid, thermal insulation, and good acoustic properties [7–9].

The first section covers the development and characteristics of NFRPCs. A brief discussion on processing techniques is presented in the second section. The third section reviews the environmental impact of NFRPCs. Potential applications of NFRPCs in lightweight constructions are highlighted in the application section. The conclusion drawn at the end closes this chapter.

8.2 BIOCOMPOSITES

A biocomposite is defined as a material developed by a matrix and reinforcement of natural fibers [10, 11]. These composites, categorized as environmentally friendly materials, are prepared using various organic or inorganic components such as natural and synthetic polymers, proteins, polysaccharides, ceramics, and metals [12]. Biocomposites are available in different shapes, such as films, membranes, coatings, particles, fibers, and foams [12].

8.2.1 Natural fiber-reinforced polymer composites (NFRPCs)

NFRPCs are among the most promising composite materials. Because of their natural fibers they have eminent properties such as low self-weight, corrosion resistance, and biodegradability [12, 13]. Thus, polymers reinforced with natural fibers are an attractive alternative for synthetic fiber-reinforced polymer composites. These biocomposites can be partly or wholly renewable and biodegradable, require minimum energy during production, and provide good physical properties for various applications [14]. One emerging application of NFRPCs is in lightweight structural composites.

Creating NFRPCs consist of two phases: the reinforcement phase and the matrix phase. Using natural fibers in the reinforcing phase improves their physical and mechanical properties and produces biodegradable and lightweight biocomposites [13, 15]. Natural fibers come from plants, animals, or minerals. Plant fibers consist of main chemical elements such as cellulose, hemicellulose, pectin, wax, and lignin [16]. The best-known natural fiber examples are stem-based fiber (flax, hemp, jute, etc.), leaf-based fibers (sisal, abaca, etc.), seed fiber (cotton, kapok), fruit (coir, palm, pineapple, etc.), grass and reed fiber (wheat, corn, and rice), and other types derived from wood and roots. An example of a structural overview of natural fiber (flax fiber) is shown in Figure 8.1.

Polymer matrices are defined as matrix phases. Generally, polymers are derived from renewable resources or fossil fuels [15]. The matrices generated from renewable resources are polylactic acid (PLA) [17], polyhydroxyalkanoates (PHA) [18], polyhydroxybutyrate (PHB) [19], polyhydroxyvalerate

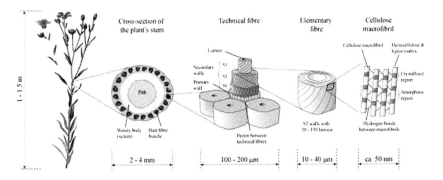

Figure 8.1 Example of detail structural overview of natural fiber (flax fiber) [24].

(PHV) [20], etc. These biopolymers are either compostable or degradable by microorganisms at the end-of-life cycle. The fossil fuel-based polymer matrices are polypropylene (PP) [21], polyethylene terephthalate (PET) [22], polycaprolactone (PCL) [23], etc. However, PCL is fully obtained from fossil fuels but is biodegradable.

Numerous studies have reported the development of various NFRPCs for lightweight applications and the evaluation of the mechanical and chemical properties, as summarized in Table 8.1. These studies analyzed the effect of the fiber dimensions, fiber content, surface modifications, and manufacturing processes on the properties of NFRPCs. Mechanical properties of NFRPCs were most commonly investigated where the mechanical performance of composites in various conditions was determined. The thermal properties, flame retardancy, and biodegradability properties were also examined for NFRPCs.

The bonding between the natural fiber and polymer influences the properties of the biocomposites. Due to the hydrophilic nature of natural fibers, they exhibit poor adhesion properties [22, 25]. Different chemical treatments on natural fibers have been used to increase interfacial properties between natural fibers and polymer matrices to overcome the limitations [6]. Techniques such as alkali treatment [26, 27], silane treatment [28–31], acetylation [32], plasma treatment [33], benzoylation treatment [34, 35], peroxide treatment [36], sodium chlorite treatment [37, 38], and fungal treatment [39, 40] were performed to modify the surface of the natural fibers. Figure 8.2 shows the surface morphology of hemp fiber with various treatments to improve the adhesion properties. As a result, after significant surface modification, the interfacial bonding of natural fibers with a polymer matrix was improved and the mechanical properties of the NFRPCs were enhanced.

8.2.2 General characteristics of NFRPCs

NFRPCs show considerable improvement in mechanical strength with natural fibers embedded in polymer matrices [14]. The mechanical strength of NFRPCs depends on the adhesion obtained between the fiber to a matrix. Essential features such as fiber orientation, moisture absorption, volume fraction, and physical properties play a crucial role in providing good mechanical properties of NFRPCs.

Haghighatnia et al. [67] developed hemp fiber-reinforced PU composites and investigated the mechanical properties of the composites. The results showed that by increasing the fiber content up to 40%, flexural strength increased by 193.24%.

Kumar et al. [68] reported the effects of fiber loading on pineapple leaf and sisal fiber-reinforced polyester composites by using the injection molding technique. The results demonstrated that maximum tensile strength, tensile modulus, and elongation of the pineapple (50 wt%) and sisal fiber

Table 8.1 List of various NFRPCs for lightweight constructions

Reinforcement phase	Matrix phase	Characterization	Processing techniques	References
Cotton	PET	Mechanical	Hand lay-up	[41]
Sisal	PET	Thermal	Twin screw extruder	[42]
Coir	PP	Mechanical, structural, hardness	Single screw extruder	[43]
Hemp	PP	Mechanical, water sorption, swelling	Blending	[44]
Coconut	PP	Mechanical, water absorption	Compounding	[45]
Bamboo	PP	Mechanical, thermal	Co-rotating twin screw extruder	[46]
Palm oil	Polyurethane (PU)	Thermal	Compression molding	[47]
Flax	PU	Thermal, mechanical	Hand lay-up	[48]
Hemp	High-density polyethylene (HDPE)	Mechanical, thermal	Injection molding	[49]
Corn fiber	PVA	Mechanical, thermal	Injection molding	[50]
Date palm	Polyvinyl alcohol	Erosion	Injection molding	[51]
Bamboo	PLA	Mechanical, thermal	Hand lay-up	[52]
Ramie	PLA	Mechanical, thermal	Co-rotating twin screw extruder	[53]
Hemp	PCL	Mechanical, thermal	Twin screw extruder	[23]
Sisal	PCL	Mechanical, thermal	Twin screw extruder	[54]
Jute	PHB	Mechanical, morphology	Hand lay-up	[55]
Hemp	PHB	Physical, mechanical, visual	Extrusion	[56]
Ramie	Polyester resin	Water absorption, density	Molding	[57]
Jute fiber	Polyester resin	Erosion	Compression molding	[58]
Jute, sisal, banana	Epoxy	Mechanical	Hand lay-up	[59]
Bagasse	Epoxy	Erosion	Hand lay-up	[60]
Bamboo fiber	Epoxy	Erosion	Hand lay-up	[61]
Kenaf	Epoxy	Wear, frictional	Pull out	[62]
Kenaf	PET and Epoxy	Mechanical, morphology, flammability	Molding	[63]
Kenaf and glass fiber	PET	Mechanical, morphology	Hand lay-up	[64]
Rice husk	PET and HDPE	Mechanical, water absorption, thickness swelling	Compression molding	[65]
Bamboo and glass fiber	PP	Mechanical, environmental	Molding	[66]

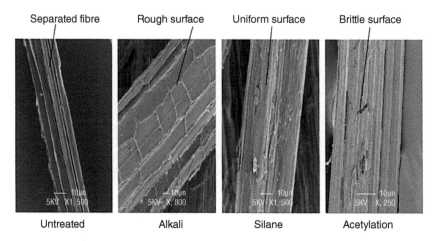

Figure 8.2 Surface morphology of the untreated and treated hemp fibers under the scanning electron microscope [31].

(50 wt%) reinforced polyester was 207.5 MPa, 4078 MPa, and 13.8%, respectively. Furthermore, it was stated that the uniform dispersion and good adhesion between the fibers and polyester matrix helped improve mechanical strength. A 15-mm fiber length was optimal for achieving the best mechanical strength.

Mohan et al. [69] evaluated the mechanical and moisture properties of the alkali-treated sida acuta-reinforced epoxy composite. The results showed that the treated sida acuta composite showed better mechanical properties than the untreated sida acuta composite.

Helaili et al. [70] studied natural fiber composites prepared from natural Alfa fibers mixed with epoxy resin 862 using the molding technique. The Alfa fibers were short and randomly distributed in the composite. The experimental results were validated with the finite element model. The results indicated that the 10% volume fraction of Alfa fibers increased the elastic modulus by about 24.16% compared to pure epoxy. The fracture stresses of the composite were obtained at 28.01 MPa in contrast to unreinforced epoxy at 14.64 MPa.

Rua et al. [71] analyzed the mechanical and flexural behavior of composites developed with fique fibers and epoxy resin. The studies stated that the mean flexural strength of 55.4 MPa was lower than that of the epoxy resin at 84.3. However, the deflection of the composite (8.9 mm) was higher than the unreinforced epoxy resin (8.3 mm). Furthermore, the composites have a non-brittle failure mode and more flexible behavior than the epoxy resin under dynamic and static loading conditions.

Komal et al. [72] studied the thermal and mechanical properties of PLA/banana fiber (PLABF) biocomposites developed through three different molding processes. The injection molding (DIM), extrusion injection molding (EIM), and extrusion compression molding (ECM) techniques were used to create composite samples. The results showed no significant influence on the hardness or thermal properties produced by these three techniques. However, the DIM and ECM biocomposites showed superior tensile (55 MPa and 60 MPa) and flexural (110 MPa and 110 MPa) properties than the ECM biocomposites (25 MPa and 50 MPa). The composites produced by EIM showed the maximum storage modulus (3863 MPa) and loss modulus (673 MPa).

Chaitanya et al. [73] studied the recyclability analysis of PLA and sisal (PLAS) biocomposites produced by the extrusion process. The biocomposites were recycled eight times, and their mechanical properties were characterized. After the third time being recycled, the PLAS composites decreased storage and loss modulus. Tensile and flexural strength was reduced by 20.9% and 21.2%, respectively. Thus, based on the characterization, the PLAS composites can be recycled up to three times.

Ramires et al. [74] prepared tannin-phenolic resin reinforced with sisal fiber (30–70 wt%) through compression molding. The analysis showed that 50 wt% of sisal fibers had a high stiffness and a lower loss modulus than pure resin.

Flame retardancy of the natural fiber-reinforced polymers can be provided using a protective surface coating for fibers. Ortega et al. [75] investigated the fire behavior of alkali and magnesium hydroxide-treated, banana fiber-reinforced polypropylene composites. The studies revealed that the surface-treated banana fiber composites showed a decrease in the propagation speed of the flame compared to untreated composites.

Naughton et al. [76] investigated the fire resistance of the hemp fiber-reinforced polyester composites, and results showed that ignition was prolonged with an increase in fiber volume by more than 25%. Additionally, fire retardance increased with an increase in fiber volume and composite thickness.

NFRPCs are mainly 60 to 70% natural fibers, and the balance are the adhesive and matrix. Principally, the degradation of the NFRPCs is primarily influenced by various microorganisms, temperature, humidity, and ultraviolet light. Degradation occurs by breaking down the chemical components of the NFRPCs [77, 78].

Karamanlioglu et al. [79] studied the effect of temperature on the degradation rate of PLA in soil, and it was observed that PLA showed rapid degradation in soil rich in microorganisms at 45°C and 50°C. Altaee et al. [80] studied the biodegradation of PHB in fertile soil, which indicated that PHB degraded to monometer and oligomers of hydroxybutyrate, which were consequently incorporated into microorganisms and their enzymatic activities. Additionally, it was mentioned that UV-treated PHB tends to decompose faster than untreated PHB samples.

Boyandin et al. [81] demonstrated biodegradation of PHA in soils with various weather conditions, and results showed that soil microorganisms heavily influence polymer chemical properties. Degradation was seen in the decrease of the molecular weight and increase of crystallinity, which refers to the potential collapse of the amorphous phase of the polymer. It was shown that microorganisms on the surface of the PHA polymer are proper destructors for disintegrating the polymers.

Gunti et al. [17] investigated the degradation properties of alkali-treated natural fiber such as jute, sisal, and elephant grass-reinforced PLA composites. The percentage of the weight loss in all the composites increased linearly with the increase of days in soil. The degradation rate was higher in the alkali-treated composites compared to untreated natural fiber composites.

Wu et al. [82] studied the biodegradability of the sisal fiber-reinforced acrylic acid-grafted polybutylene terephthalate (PBT-g-AA/SF). The polybutylene terephthalate (PBT) and PBT-g-AA/SF were buried in the soil for degradation tests. The rate of degradation was slower in pure PBT compared to PBT-g-AA/SF composites, and the rate of degradation of PBT-g-AA/SF was greater than 50%.

Chaudhuri et al. [83] studied jute-reinforced HDPE composite biodegradation in soil and pure microbial culture. The composites were exposed to both conditions for two to six months, and results indicated that the maximum jute-reinforced HDPE showed ultimate biodegradation in both treatment media after six months. The biodegradation process was significantly slower in microbial media compared with soil.

Fakhrul et al. [84] studied the degradation behavior of PP reinforced with sawdust and wheat flour. The PP with natural fibers was exposed to soil, water, and a brine solution in an open atmosphere for 15 days. The results showed that the addition of natural fibers enhanced the biodegradability of PP. However, samples exposed to water showed an average degradation compared to soil and brine solution. The PP sawdust samples revealed a better degradation rate than PP with wheat flour because of the increased high-water absorption.

Briese et al. [85] studied the degradation rate of PHB in sewage sludge for 12 weeks. It was discovered that brittle bottle pieces of PHB decomposed entirely over the period.

Breslin et al. [86] studied the starch polymer composite degradation in municipal solid waste landfills, and it was found that starch polymer composites weakened faster than pure polymer over two years in a landfill. Thus, it was evident that natural fiber-reinforced polymers deposited in landfills decompose slowly.

Research groups developed various biodegradability assessments under different environmental conditions [87, 88]. However, there is a lack of a single standard technique for evaluating the biodegradability of biocomposites. Moreover, standard methods of analyzing biodegradability need to be developed to use environmentally friendly composites as an alternative to synthetic fiber composites.

8.3 PROCESSING TECHNIQUES

Generally, various processing techniques have been developed to produce NFRPCs according to their applications [16, 22]. The most frequently used processes are hand lay-up [89–91], filament winding [92], vacuum resin molding [93], pultrusion [94], injection molding [95–97], and compression molding [95]. Additionally, the processing techniques also depend on the fiber length (short or long) and orientation (continuous or discontinuous) [96, 98]. For instance, short fibers are more appropriate for injection molding and large fibers for compression molding [98]. The filament winding and compression molding techniques are suitable for long fibers larger than 3 mm. However, processing parameters are also selected based on the manufacturing cost, production time, and performance of the products. According to Abraham et al. [99], vacuum infusion methods are superior in obtaining biocomposites without voids, contamination, and accurate fiber spacing compared to the hand lay-up technique.

8.3.1 Hand lay-up

The hand lay-up technique (Figure 8.3) is a manual and easy molding technique used for producing biocomposites [89–91], which involves several steps. Initially, all the mold surface is cured by an anti-adhesive realizing agent or Teflon paper coated with wax to prevent the polymer resin or matrix from sticking to the surface of the mold [89–91]. Then, a thin layer of polymer matrix or resin is poured or sprayed at the top and bottom of the mold plate to get a smooth surface. The layers of fibers are cut into required shapes and placed on the surface of the mold. The resin with other elements is used to brush the fibers to disperse uniformly in the mold. Afterwards, the laminated layers are squeezed using a roller to remove trapped air bubbles and excess polymer [89–91]. The mold is closed for a certain period, such as 24 hours, to ensure hardening into a single composite sheet. Thermoplastic polymers such as epoxy [100], polyester [101], and polyurethane [102] are primarily used for hand lay-up techniques.

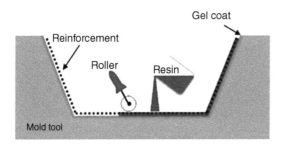

Figure 8.3 Illustration of hand lay-up [103].

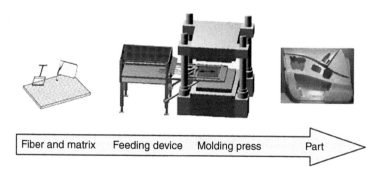

Figure 8.4 Configuration of compression molding technique [12].

8.3.2 Compression molding

Compression molding is an old and bulk scale processing technique for manufacturing biocomposites [16, 104–106]. The method is usually used for thermoplastic matrices with chopped fibers or mats of long and short fibers. The fibers are aligned randomly or control positioned within the matrices [98, 104, 105]. The illustration of compression molding is shown in Figure 8.4. The composites are produced through hot pressing by placing the materials in the open space mold cavity. Developing composites through compression molding avoids the requirements of bulk raw materials at the early stage of the product development.

The process involves multiple steps. Initially, the material is placed in hot mold, and the mold is closed by hydraulic press. The mold is pressed under high pressure and temperature. The polymeric materials are liquefied through heat at a constant pressure to fill the mold [98, 104, 105]. The mold is then cooled to room temperature before removing the pressure to obtain a final product. However, the cooling phase is not essential for thermoset polymers, as they harden under applied pressure. The vital parameters to maintain during the process are thickness, time, polymer melting temperature, and pressure to obtain a good biocomposite product [98].

8.3.3 Injection molding

The injection molding technique is used in the batch production of products in manufacturing companies [97]. The injection molding equipment is shown in Figure 8.5. The process includes three zones such as the feeding zone, transition zone, and injection zone. The procedure begins by feeding fiber and polymer into the hopper connected to the nozzle. The nozzle is heated to the desired melting temperature of the polymer. Then, the liquefied material is injected into a closed mold with the preferred shape through a force by screw. Then the screw is retracted, and the mold is cooled to room temperature to obtain the finished product.

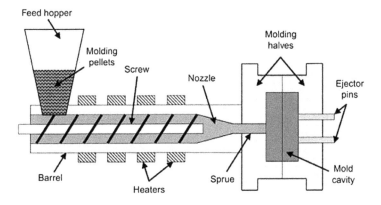

Figure 8.5 Injection molding [109].

Short fibers are used in this randomly oriented technique [98, 107]. However, certain fibers can be aligned depending on the matrix viscosity and mold design [107]. Siva et al. [108] investigated the difference in the properties of a thermoplastic hemp fiber composite prepared by injection molding and extrusion-injection molding. The results showed that the extrusion-injection molding process enhanced the bonding between fiber and matrix compared to injection molding. The adhesion was obtained through high-intensity mixing achieved through a twin-screw extruder. The voids in the produced composite decreased compared to injection molding.

8.3.4 Resin transfer molding (RTM)

RTM is typically a closed molding manufacturing process where the resin is transferred over already placed reinforcement fibers [16, 22, 110]. Fibers are placed on the mold cavity, the mold is closed with clamps, and resin is infused into the mold cavity. The injection of resin is pumped through a vented mold with pressure. The resin infusion stops once the cavity is full, and the resin is left to solidify before demolding. Temperature, resin viscosity, mold structure, and injection pressure are vital in obtaining good-quality composite products. This process includes lower temperature requirements than other methods and aids in the thermal degradation of materials [16, 111].

8.3.5 Vacuum-assisted resin transfer molding (VARTM)

VARTM is a modification of the RTM process [112, 113]. The main advantage of VARTM is that it is low-cost and provides a good-quality final product. VARTM uses the excellent fiber-to-resin ratio control and provides consistent manufacturing products for bulk fabrication.

The process involves placing dry fiber mats into a mold and covering it with a flexible vacuum bag, as shown in Figure 8.6. The vacuum bag is

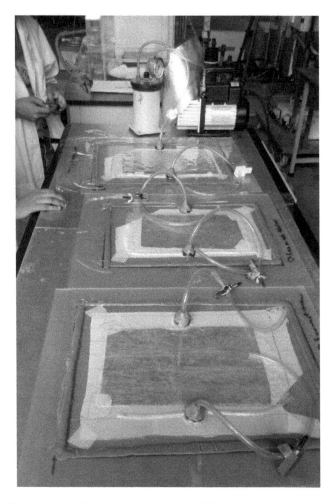

Figure 8.6 Vacuum-assisted resin infusion technique [93].

firmly sealed with adhesive tape, then the air is forced out of the bag to create a vacuum. At the same time, a polymer resin is infused to reinforce the
dry fibers until the resin is completely treated. Finally, the laminated composite is removed with the aid of peel ply. The process can be problematic if
the vent line location and injection line are not designed or aligned accurately to fill the resin in the mold cavity [112, 114].

8.3.6 Filament winding

Filament winding is a low-cost manufacturing process for producing composite structures. The method of filament winding is shown in Figure 8.7.
The process is a continuous reinforcement technique in which fibers are

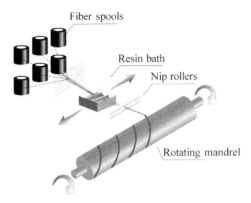

Figure 8.7 Filament winding technique [116].

passed through a resin bath and coiled onto a rotating mandrel [115], which is computer-controlled and automated. The fiber strands or tows pass through the resin bath in controllable fiber orientation. The fibers are impregnated with resin in the bath, and the impregnated strands are wound. The angle of fibers can be based on the product manufacturing requirements. Resin temperature, winding speed, and fiber tension are essential to obtain excellent-quality biocomposite products.

8.3.7 Pultrusion

Pultrusion involves continuous fibers infused with the polymer matrix using a heated mold to form composites [94]. The technique shown in Figure 8.8 facilitates obtaining constant cross-sectional profiles for batch production compared to other processing techniques [117]. The composites created from pultrusion have approximately 70% fiber content [117, 118] and are firm. Due to its straightforward high quality and low cost, pultrusion is indispensable for creating natural fiber composite products.

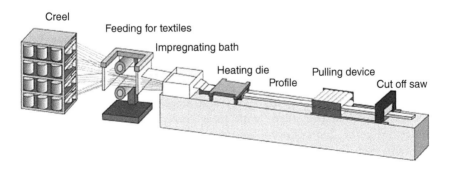

Figure 8.8 Illustration of pultrusion equipment [12].

The process starts with the fibers being hauled from the holder into a resin bath. The fibers are wholly immersed in the polymer resin in the bath to avoid insufficient fiber coating or wetting. Then, the fiber guide plate aligns most of the fiber horizontally before entering into a heated mold. Simultaneously, the mold is heated using an aluminum heater block attached with a thermocouple with well-regulated temperature controls. The heat inside the mold is distributed into two zones: the gelation and curing zone [119].

Once the fibers enter the mold, gelation of the resin begins. Afterwards, the composite is cured and removed from the mold using a puller. Temperature, amount of resin, and pulling speed are critical factors in obtaining good quality composite fibers. The temperature during the process ranges between 100 and 200°C. Additionally, the pultruded combined size also affects the pulling speed and temperature.

8.4 IMPACT OF ENVIRONMENTAL PARAMETERS ON NFRPCs

Despite the advantages of NFRPCs, the use of biocomposites for peripheral applications is limited, owing to the degradation risk when exposed to different environmental conditions [120]. Harsh environmental conditions such as humidity [121], temperature [122], high energy radiation [123], and weathering conditions [124] cause NFRPCs to deteriorate.

Studies have been [125–127] conducted to define the resistance of NFRPCs in environmental conditions and to analyze the rate of weakening of materials. However, the analysis couldn't forecast the end of product life. For example, artificial aging testing cabinets are used to predict the degradation of products exposed to weathering conditions.

Badji et al. [128] described the relationship between artificial and natural weathering of PP reinforced with hemp fiber biocomposites. It was reported that after 250 h of exposure to two weathering conditions, pure PP showed similar deterioration to that of one year of natural weathering conditions. On the contrary, the natural fiber biocomposites showed the same degradation rate in both states after 750 h.

Dayo et al. [121] demonstrated the weathering effects (25°–30°C temperature and 40%–80% relative humidity) on the mechanical properties of hemp fiber-treated polybenzoxazine composites. The results indicated that the moisture absorption increased, and the mechanical properties decreased with increasing temperature and relative humidity.

Chee et al. [129] investigated the influence of UV radiation at higher temperatures on kenaf-reinforced epoxy and bamboo-reinforced epoxy composites and compared them with pure epoxy. The study showed that noticeable color changes occurred after exposing the composites to higher UV radiation; the increase in fiber content caused the color changes. The composites remain unchanged in terms of their thermal stability and mechanical properties after exposure to accelerated conditions.

Yu et al. [130] investigated the starch-reinforced PLA biocomposites aging in UV radiation. The results showed that biocomposite tensile strength increased after exposure for a short period, but the strength gradually decreased with increased radiation time.

Assarar et al. [131] analyzed water aging on the mechanical properties of flax fiber-reinforced epoxy composites. The results indicated that water aging significantly decreased both the modulus and maximum strain of the flax fiber composite; water aging did not affect elastic properties. In another study, sisal fiber-reinforced PP composites were subjected to hot water immersion to study the effects of moisture absorption [132]. The analysis showed that the tensile strength of the composites constantly decreased with an increase in immersion time.

The primary concern regarding exterior applications of NFRPCs is in their dimensional stability and durability against environmental parameters. The degradation of physical and mechanical properties of NFRPCs upon long-term exposure ecological conditions is a major limiting factor to their application. Thus, surface modifications of natural fibers that improve their resistance is essential for the development of resistant and sustainable materials.

8.5 APPLICATIONS

The properties of biocomposites are analogous to traditional fiber composites, but they offer additional biodegradable and renewable properties. Biocomposites have been widely used in lightweight construction applications such as construction, transport, marine, and sports over the previous few years because of their excellent behavioral properties such as corrosion resistance, low processing energy requirements, and some essential specific properties. Several factors impact the applications of biocomposites, including durability, moisture resistance, and thermal stability.

8.5.1 Automotive

In the automobile industry, NFRPCs possess various advantages such as low cost, reduced weight, improved acoustic insulation, and increased life expectancy and eco-friendliness compared to conventional composites [133]. Automobile body parts such as dashboards, engine hoods, and storage tanks are manufactured using NFRPCs, as shown in Figure 8.9 [133, 134]. The main advantage with NFRPCs products is that they reduce overall weight and decrease carbon emissions. Coir fibers first gained popularity as back pads, head restraints, and seat bottoms [135, 136]. Then, major companies like Ford, Honda, General Motors, and Volkswagen started using natural fiber (jute, hemp, and kenaf) reinforced composites for interior products manufacturing [133, 137–139].

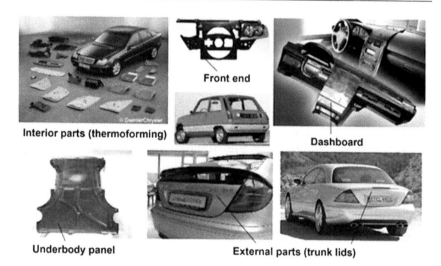

Figure 8.9 Fiber properties and their recent application in the automotive industry [142].

Hemp fiber-reinforced polyester composites were used in the "Lotus Eco Elise" concept car [140]. The processing techniques used were hand lay-up and vacuum bagging to form the parts. Changing the material from glass fiber to hemp fiber resulted in reduced weight, higher fuel efficiency, and better car performance than other vehicles in that range [140].

A natural fiber mat was used in a door panel for the BMW 7 Series model, as shown in Figure 8.9 [141]. The panels were developed using the compression molding technique with thermosetting acrylic copolymer and prepreg natural fibers. Case studies indicated that natural fiber-reinforcement facilitates the manufacture of composites with high loading. Additionally, carbon emission is decreased without forfeiting vehicle performance. Toyota, a Japanese automobile manufacturing company, used kenaf-reinforced PLA in its RAUM 2003 model to fabricate spare tire covers [88]. Toyota also later developed soy-based seat foams in 2008 RAV-4 models [88].

8.5.2 Aerospace

NFRPCs were first used in the automotive industry and then later adopted for structural designs in the aerospace industry [143]. Based on airbus composite training, approximately 52% of the structural weight is designed by composite materials [144]. The Council for Scientific and Industrial Research, South Africa, developed NFRPCs with flax-reinforced phenolic resin to produce panels for cabin interiors [145]; Boeing Research and Technology manufactured the panels using recyclable thermoplastic reinforced with flax fibers [146]. The NFRPC products are 35% lighter than carbon fiber-reinforced epoxy prepregs. Flax fiber biocomposites were also used for cabin sidewalls [146].

A study examined the aircraft wing box made using ramie fiber biocomposites, and results showed that component weight was reduced, compared to aluminum alloy components [143]. Eloy et al. [147] reported the possible application of NFRPCs for aerospace interiors parts using cotton fiber-reinforced polyurethane. The studies stated that surface modifications of natural fiber and that PU reinforced with cotton fiber is feasible for aircraft components.

8.5.3 Constructions

Most building components are manufactured using conventional materials such as steel, bricks, and cement [148], which consume a lot of thermal and electrical energy. Over the past few years, there has been a significant shift towards replacing traditional constructional materials with lightweight building materials, as shown in Figure 8.10 [149]. These lightweight building materials are produced mainly by natural fiber-reinforced degradable or renewable polymer composites [150, 151]. These offer excellent benefits as

Figure 8.10 Natural fiber composites for construction [148].

building materials, as they are renewable materials with a lower cost, are bio-degradable, and have good mechanical and acoustic properties [151, 152].

Rice husk-reinforced phenolic resin composites products have been used to fabricate particleboards for building temporary shelters [153]. The boards are used as partition sheets, ceilings, and window frames, which reduces operational costs and requires less construction material [153]. A banana fiber-reinforced epoxy composite was used to design and fabricate a household furniture application as an alternative to metallic and non-metallic materials [153]. Li et al. (Li, Wang, and Wang 2006) reported the coir fiber-reinforcement cementitious effect improved the flexural strength more than 10 times. In structural applications, natural fiber reinforced with PP and PE has been widely used in decking [154], doors [155], insulating wall frames [156], floor laminates [157], and window frames [158]. The results showed that NFRPCs could replace conventional steel-reinforced concrete structures and make the building industry more environmentally friendly.

8.5.4 Other applications

Flax fiber combined with carbon fiber [159] is used to manufacture bicycle frames. This hybridization has resulted in a similar strength and stiffness as commercially available carbon, aluminum, and titanium frames, while also possessing better damping properties [159]. These extraordinary properties were achieved by maintaining 40 wt% bio-content [159]. The vibration damping of materials where bidirectional flax is used has superior damping properties than other materials. Sicomin Epoxy Systems (France) has manufactured hull, deck, and interior components with flax fiber-reinforced GreenPoxy InfuGreen 810 resin for oceangoing vessels. Naval architects Judel/Vrolijk & Co. designed and promoted 100% renewable and eco-friendly materials without compromising performance [160].

8.6 CONCLUSION

Natural fibers are emerging alternatives utilized in composite materials. The advantages of using natural fibers are high availability, low cost, light weight, high strength-to-weight ratio, and they are biodegradable and nonhazardous. Researchers have worked to integrate NFRPCs in various capacities—and one promising application is lightweight structural materials due to low self-weight. As a lightweight material, NFRPCs are ubiquitous in the automobile, aerospace, and building industries. Poor interfacial bonding and the environmental impact of increasing product life expectancy limits the advancement of these materials; however, with progress in research and development, there is a good possibility for natural fiber polymer composites as alternative materials in lightweight structural applications and in the development of environmentally friendly materials.

REFERENCES

1. H. N. Dhakal, S. O. Ismail, Lightweight composites, important properties and applications, in: *Sustainable Composites for Lightweight Applications*, Elsevier, 2021: pp. 53–119. https://doi.org/10.1016/B978-0-12-818316-8.00006-2
2. Y. Sun, Y. Chu, W. Wu, H. Xiao, Nanocellulose-based lightweight porous materials: A review, *Carbohydr. Polym.* 255 (2021) 117489. https://doi.org/10.1016/j.carbpol.2020.117489
3. A. K. Mohanty, S. Vivekanandhan, J.-M. Pin, M. Misra, Composites from renewable and sustainable resources: Challenges and innovations, *Science* 362 (2018) 536–542. https://doi.org/10.1126/science.aat9072
4. P. Chatterjee, N. Mandal, S. Dhar, S. Chatterjee, S. Chakraborty, A novel decision-making approach for light weight environment friendly material selection, *Mater. Today Proc.* 22 (2020) 1460–1469. https://doi.org/10.1016/j.matpr.2020.01.504
5. G. S. Cole, A. M. Sherman, Light weight materials for automotive applications, *Mater. Charact.* 35 (1995) 3–9. https://doi.org/10.1016/1044-5803(95)00063-1
6. O. Faruk, A. K. Bledzki, H.-P. Fink, M. Sain, Progress report on natural fiber reinforced composites, *Macromol. Mater. Eng.* 299 (2014) 9–26. https://doi.org/10.1002/mame.201300008
7. A. Bhat, J. Naveen, M. Jawaid, M. N. F. Norrrahim, A. Rashedi, A. Khan, Advancement in fiber reinforced polymer, metal alloys and multi-layered armour systems for ballistic applications – A Review, *J. Mater. Res. Technol.* (2021). https://doi.org/10.1016/j.jmrt.2021.08.150
8. A. Behera, B. Swain, D. K. Sahoo, Fiber-reinforced ceramic matrix nanocomposites, in: *Fiber-Reinforced Nanocomposites Fundam Application*, Elsevier, 2020: pp. 359–368. https://doi.org/10.1016/B978-0-12-819904-6.00016-5
9. R. Reshmy, E. Philip, P. H. Vaisakh, R. Sindhu, P. Binod, A. Madhavan, A. Pandey, R. Sirohi, A. Tarafdar, Biodegradable polymer composites, in: *Biomass, Biofuels, Biochem*, Elsevier, 2021: pp. 393–412. https://doi.org/10.1016/B978-0-12-821888-4.00003-4
10. A. K. Mohanty, M. Misra, L. T. Drzal, eds., *Natural Fibers, Biopolymers, and Biocomposites*, CRC Press, 2005. https://doi.org/10.1201/9780203508206
11. K. Haraguchi, Biocomposites, in: *Encycl. Polym. Nanomater.*, Springer Berlin Heidelberg, Berlin, Heidelberg, 2014, pp. 1–8. https://doi.org/10.1007/978-3-642-36199-9_316-1
12. U. Riedel, Biocomposites, in: *Polym. Sci. A Compr. Ref.*, Elsevier, 2012: pp. 295–315. https://doi.org/10.1016/B978-0-444-53349-4.00268-5
13. L. Averous, N. Boquillon, Biocomposites based on plasticized starch: Thermal and mechanical behaviours, *Carbohydr. Polym.* 56 (2004) 111–122. https://doi.org/10.1016/j.carbpol.2003.11.015
14. H. Ku, H. Wang, N. Pattarachaiyakoop, M. Trada, A review on the tensile properties of natural fiber reinforced polymer composites, *Compos. Part B Eng.* 42 (2011) 856–873. https://doi.org/10.1016/j.compositesb.2011.01.010
15. M. Nagalakshmaiah, S. Afrin, R. P. Malladi, S. Elkoun, M. Robert, M. A. Ansari, A. Svedberg, Z. Karim, Biocomposites, in: *Green Compos. Automot. Appl.*, Elsevier, 2019: pp. 197–215. https://doi.org/10.1016/B978-0-08-102177-4.00009-4

16. S. Mahmud, K. M. F. Hasan, M. A. Jahid, K. Mohiuddin, R. Zhang, J. Zhu, Comprehensive review on plant fiber-reinforced polymeric bio-composites, *J. Mater. Sci.* 56 (2021) 7231–7264. https://doi.org/10.1007/s10853-021-05774-9

17. R. Gunti, A. V. Ratna Prasad, A. V. S. S. K. S. Gupta, Mechanical and degradation properties of natural fiber-reinforced PLA composites: Jute, sisal, and elephant grass, *Polym. Compos.* 39 (2018) 1125–1136. https://doi.org/10.1002/pc.24041

18. E. Ten, L. Jiang, J. Zhang, M. P. Wolcott, Mechanical performance of poly-hydroxyalkanoate (PHA)-based biocomposites, in: *Biocomposites*, Elsevier, 2015: pp. 39–52. https://doi.org/10.1016/B978-1-78242-373-7.00008-1

19. M. K. M. Smith, D. M. Paleri, M. Abdelwahab, D. F. Mielewski, M. Misra, A. K. Mohanty, Sustainable composites from poly(3-hydroxybutyrate) (PHB) bioplastic and agave natural fibre, *Green Chem.* 22 (2020) 3906–3916. https://doi.org/10.1039/D0GC00365D

20. R. A. Shanks, A. Hodzic, S. Wong, Thermoplastic biopolyester natural fiber composites, *J. Appl. Polym. Sci.* 91 (2004) 2114–2121. https://doi.org/10.1002/app.13289

21. P. Luthra, K. K. Vimal, V. Goel, R. Singh, G. S. Kapur, Biodegradation studies of polypropylene/natural fiber composites, *SN Appl. Sci.* 2 (2020) 512. https://doi.org/10.1007/s42452-020-2287-1

22. V. Chauhan, T. Kärki, J. Varis, Review of natural fiber-reinforced engineering plastic composites, their applications in the transportation sector and processing techniques, *J. Thermoplast. Compos. Mater.* (2019). https://doi.org/10.1177/0892705719889095

23. H. N. Dhakal, S. O. Ismail, J. Beaugrand, Z. Zhang, J. Zekonyte, Characterization of nano-mechanical, surface and thermal properties of hemp fiber-reinforced polycaprolactone (HF/PCL) Biocomposites, *Appl. Sci.* 10 (2020) 2636. https://doi.org/10.3390/app10072636

24. W. Woigk, C. A. Fuentes, J. Rion, D. Hegemann, A. W. van Vuure, C. Dransfeld, K. Masania, Interface properties and their effect on the mechanical performance of flax fibre thermoplastic composites, *Compos. Part A Appl. Sci. Manuf.* 122 (2019) 8–17. https://doi.org/10.1016/j.compositesa.2019.04.015

25. K. Roy, S. C. Debnath, A. Pongwisuthiruchte, P. Potiyaraj, Recent advances of natural fibers based green rubber composites: Properties, current status, and future perspectives, *J. Appl. Polym. Sci.* 138 (2021) 50866. https://doi.org/10.1002/app.50866

26. M. M. Kabir, H. Wang, K. T. Lau, F. Cardona, Chemical treatments on plant-based natural fibre reinforced polymer composites: An overview, *Compos. Part B Eng.* 43 (2012) 2883–2892. https://doi.org/10.1016/j.compositesb.2012.04.053

27. W. Liu, A. K. Mohanty, L. T. Drzal, P. Askel, M. Misra, Effects of alkali treatment on the structure, morphology and thermal properties of native grass fibers as reinforcements for polymer matrix composites, *J. Mater. Sci.* 39 (2004) 1051–1054. https://doi.org/10.1023/B:JMSC.0000012942.83614.75

28. M. Asim, M. Jawaid, K. Abdan, M. R. Ishak, Effect of alkali and silane treatments on mechanical and fibre-matrix bond strength of kenaf and pineapple leaf fibres, *J. Bionic Eng.* 13 (2016) 426–435. https://doi.org/10.1016/S1672-6529(16)60315-3

29. Y. Xie, C. A. S. Hill, Z. Xiao, H. Militz, C. Mai, Silane coupling agents used for natural fiber/polymer composites: A review, *Compos. Part A Appl. Sci. Manuf.* 41 (2010) 806–819. https://doi.org/10.1016/j.compositesa.2010.03.005

30. A. Atiqah, M. Jawaid, M. R. Ishak, S. M. Sapuan, Effect of alkali and silane treatments on mechanical and interfacial bonding strength of sugar palm fibers with thermoplastic polyurethane, *J. Nat. Fibers.* 15 (2018) 251–261. https://doi.org/10.1080/15440478.2017.1325427

31. M. M. Kabir, H. Wang, K. T. Lau, F. Cardona, T. Aravinthan, Mechanical properties of chemically-treated hemp fibre reinforced sandwich composites, *Compos. Part B Eng.* 43 (2012) 159–169. https://doi.org/10.1016/j.compositesb.2011.06.003

32. I. Mukhtar, Z. Leman, E. S. Zainudin, M. R. Ishak, Effectiveness of alkali and sodium bicarbonate treatments on sugar palm fiber: Mechanical, thermal, and chemical investigations, *J. Nat. Fibers.* 17 (2020) 877–889. https://doi.org/10.1080/15440478.2018.1537872

33. N. A. Ibrahim, B. M. Eid, M. S. Abdel-Aziz, Effect of plasma superficial treatments on antibacterial functionalization and coloration of cellulosic fabrics, *Appl. Surf. Sci.* 392 (2017) 1126–1133. https://doi.org/10.1016/j.apsusc.2016.09.141

34. S. Mohd Izwan, S. M. Sapuan, M. Y. M. Zuhri, A. R. Mohamed, Effects of benzoyl treatment on NaOH treated sugar palm fiber: Tensile, thermal, and morphological properties, *J. Mater. Res. Technol.* 9 (2020) 5805–5814. https://doi.org/10.1016/j.jmrt.2020.03.105

35. R. Abdul Majid, H. Ismail, R. Mat Taib, Processing, tensile, and thermal studies of Poly(Vinyl Chloride)/epoxidized natural rubber/kenaf core powder composites with benzoyl chloride treatment, *Polym. Plast. Technol. Eng.* 57 (2018) 1507–1517. https://doi.org/10.1080/03602559.2016.1211687

36. N. Razak, N. Ibrahim, N. Zainuddin, M. Rayung, W. Saad, The influence of chemical surface modification of kenaf fiber using hydrogen peroxide on the mechanical properties of biodegradable kenaf fiber/Poly(Lactic Acid) composites, *Molecules.* 19 (2014) 2957–2968. https://doi.org/10.3390/molecules19032957

37. V. G. Geethamma, K. Thomas Mathew, R. Lakshminarayanan, S. Thomas, Composite of short coir fibres and natural rubber: Effect of chemical modification, loading and orientation of fibre, *Polymer (Guildf).* 39 (1998) 1483–1491. https://doi.org/10.1016/S0032-3861(97)00422-9

38. J. B. Reeves, Sodium chlorite treatment of plant materials: Fiber and lignin composition, digestibility, and their interrelationships, *J. Dairy Sci.* 70 (1987) 2534–2549. https://doi.org/10.3168/jds.S0022-0302(87)80322-3

39. D. Gulati, M. Sain, Fungal-modification of Natural Fibers: A novel method of treating natural fibers for composite reinforcement, *J. Polym. Environ.* 14 (2006) 347–352. https://doi.org/10.1007/s10924-006-0030-7

40. M. Y. Khalid, R. Imran, Z. U. Arif, N. Akram, H. Arshad, A. Al Rashid, F. P. García Márquez, developments in chemical treatments, manufacturing techniques and potential applications of natural-fibers-based biodegradable composites, *Coatings.* 11 (2021) 293. https://doi.org/10.3390/coatings11030293

41. Y. Yao, M. Li, M. Lackner, L. Herfried, A continuous fiber-reinforced additive manufacturing processing based on PET fiber and PLA, *Materials (Basel).* 13 (2020) 3044. https://doi.org/10.3390/ma13143044

42. A. D. Gudayu, L. Steuernagel, D. Meiners, R. Gideon, Characterization of the dynamic mechanical properties of sisal fiber reinforced PET composites; Effect of fiber loading and fiber surface modification, *Polym. Polym. Compos.* (2021). https://doi.org/10.1177/09673911211023032

43. M. M. Haque, M. Hasan, M. S. Islam, M. E. Ali, Physico-mechanical properties of chemically treated palm and coir fiber reinforced polypropylene composites, *Bioresour. Technol.* 100 (2009) 4903–4906. https://doi.org/10.1016/j.biortech.2009.04.072

44. H. Hargitai, I. Rácz, R. D. Anandjiwala, Development of HEMP fiber reinforced polypropylene composites, *J. Thermoplast. Compos. Mater.* 21 (2008) 165–174. https://doi.org/10.1177/0892705707083949

45. K. Shahril, A. Nizam, M. Sabri, A. S. Rohana, H. Salmah, Effect of chemical modifier on the properties of polypropylene (PP)/coconut fiber (CF) in automotive application, (2016). https://doi.org/10.5281/ZENODO.1127898

46. Nguyen Tri Phuong, C. Sollogoub, A. Guinault, Relationship between fiber chemical treatment and properties of recycled pp/bamboo fiber composites, *J. Reinf. Plast. Compos.* 29 (2010) 3244–3256. https://doi.org/10.1177/0731684410370905

47. R. Ghazali, Preliminary study on microbial degradation of flexible polyurethane foams-physico-mechanical and weight changes during fungal deterioration, *J. Oil Palm Res.* 17 (2005). 103-109.

48. J. Gassan, T. Dietz, A. K. Bledzki, Effect of silicone interphase on the mechanical properties of flax-polyurethane composites, *Compos. Interfaces.* 7 (2000) 103–115. https://doi.org/10.1163/156855400300184262

49. S. Singh, D. Deepak, L. Aggarwal, V. K. Gupta, Tensile and flexural behavior of hemp fiber reinforced virgin-recycled HDPE matrix composites, *Procedia. Mater. Sci.* 6 (2014) 1696–1702. https://doi.org/10.1016/j.mspro.2014.07.155

50. P. Cinelli, E. Chiellini, J. W. Lawton, S. H. Imam, Properties of injection molded composites containing corn fiber and Poly(Vinyl Alcohol), *J. Polym. Res.* 13 (2006) 107–113. https://doi.org/10.1007/s10965-005-9012-z

51. J. R. Mohanty, S. N. Das, H. C. Das, T. K. Mahanta, S. B. Ghadei, Solid particle erosion of date palm leaf fiber reinforced polyvinyl alcohol composites, *Adv. Tribol.* 2014 (2014) 1–8. https://doi.org/10.1155/2014/293953

52. R. Sukmawan, H. Takagi, A. N. Nakagaito, Strength evaluation of cross-ply green composite laminates reinforced by bamboo fiber, *Compos. Part B Eng.* 84 (2016) 9–16. https://doi.org/10.1016/J.COMPOSITESB.2015.08.072

53. T. Yu, N. Jiang, Y. Li, Study on short ramie fiber/poly(lactic acid) composites compatibilized by maleic anhydride, *Compos. Part A Appl. Sci. Manuf.* 64 (2014) 139–146. https://doi.org/10.1016/j.compositesa.2014.05.008

54. V. P. Cyras, S. Iannace, J. M. Kenny, A. Vázquez, Relationship between processing and properties of biodegradable composites based on PCL/starch matrix and sisal fibers, *Polym. Compos.* 22 (2001) 104–110. https://doi.org/10.1002/pc.10522

55. E. Rodríguez, G. Francucci, PHB coating on jute fibers and its effect on natural fiber composites performance, *J. Compos. Mater.* 50 (2016) 2047–2058. https://doi.org/10.1177/0021998315601203

56. A. T. Michel, S. L. Billington, Characterization of poly-hydroxybutyrate films and hemp fiber reinforced composites exposed to accelerated weathering, *Polym. Degrad. Stab.* 97 (2012) 870–878. https://doi.org/10.1016/j.polymdegradstab.2012.03.040

57. W. Lei, W. Lei, C. Ren, Effect of volume fraction of ramie cloth on physical and mechanical properties of ramie cloth/UP resin composite, *Trans. Nonferrous Met. Soc. China.* 16 (2006) s474–s477. https://doi.org/10.1016/S1003-6326(06)60237-9

58. G. Kalusuraman, S. Thirumalai Kumaran, M. Aslan, T. Küçükömeroğluc, I. Siva, Use of waste copper slag filled jute fiber reinforced composites for effective erosion prevention, *Measurement.* 148 (2019) 106950. https://doi.org/10.1016/j.measurement.2019.106950

59. J. I. Preet Singh, V. Dhawan, S. Singh, K. Jangid, Study of effect of surface treatment on mechanical properties of natural fiber reinforced composites, *Mater. Today Proc.* 4 (2017) 2793–2799. https://doi.org/10.1016/J.MATPR.2017.02.158

60. P. Mishra, S. K. Acharya, Solid particle erosion of Bagasse fiber reinforced epoxy composite, *Int. J. Phys. Sci.* 5 (2010) 109–115.

61. H. Jena, A. K. Pradhan, M. K. Pandit, Study of solid particle erosion wear behavior of bamboo fiber reinforced polymer composite with cenosphere filler, *Adv. Polym. Technol.* 37 (2018) 761–769. https://doi.org/10.1002/adv.21718

62. C. W. Chin, B. F. Yousif, Potential of kenaf fibres as reinforcement for tribological applications, *Wear.* 267 (2009) 1550–1557. https://doi.org/10.1016/j.wear.2009.06.002

63. M. J. Suriani, H. A. Zainudin, R. A. Ilyas, M. Petrů, S. M. Sapuan, C. M. Ruzaidi, R. Mustapha, Kenaf fiber/pet yarn reinforced epoxy hybrid polymer composites: Morphological, tensile, and flammability properties, *Polymers (Basel).* 13 (2021) 1532. https://doi.org/10.3390/polym13091532

64. M. Amir, R. Irmawaty, M. Hustim, I. R. Rahim, Tensile strength of glass fiber-reinforced waste PET and Kenauf hybrid composites, *IOP Conf. Ser. Earth Environ. Sci.* 419 (2020) 12061. https://doi.org/10.1088/1755-1315/419/1/012061

65. R. S. Chen, S. Ahmad, S. Gan, Rice husk bio-filler reinforced polymer blends of recycled HDPE/PET: Three-dimensional stability under water immersion and mechanical performance, *Polym. Compos.* 39 (2018) 2695–2704. https://doi.org/10.1002/pc.24260

66. M. M. Thwe, K. Liao, Effects of environmental aging on the mechanical properties of bamboo-glass fiber reinforced polymer matrix hybrid composites, *Compos. Part A Appl. Sci. Manuf.* 33 (2002) 43–52. https://doi.org/10.1016/S1359-835X(01)00071-9

67. T. Haghighatnia, A. Abbasian, J. Morshedian, Hemp fiber reinforced thermoplastic polyurethane composite: An investigation in mechanical properties, *Ind. Crops Prod.* 108 (2017) 853–863. https://doi.org/10.1016/j.indcrop.2017.07.020

68. S. Sathees Kumar, R. Muthalagu, C. Nithin Chakravarthy, Effects of fiber loading on mechanical characterization of pineapple leaf and sisal fibers reinforced polyester composites for various applications, *Mater. Today Proc.* 44 (2021) 546–553. https://doi.org/10.1016/j.matpr.2020.10.214

69. H. K. Chandra Mohan, S. Devaraj, R. Ranganatha, K. S. Narayana Swamy, Mechanical and moisture absorption behaviour of alkali treated sida acuta composite, *Trans. Indian Inst. Met.* (2021). https://doi.org/10.1007/s12666-021-02360-0

70. S. Helaili, M. Chafra, Y. Chevalier, Natural fiber alfa/epoxy randomly reinforced composite mechanical properties identification, *Structures.* 34 (2021) 542–549. https://doi.org/10.1016/j.istruc.2021.07.095

71. J. Rua, M. F. Buchely, S. N. Monteiro, G. I. Echeverri, H. A. Colorado, Impact behavior of laminated composites built with fique fibers and epoxy resin: a mechanical analysis using impact and flexural behavior, *J. Mater. Res. Technol.* 14 (2021) 428–438. https://doi.org/10.1016/j.jmrt.2021.06.068

72. U. K. Komal, M. K. Lila, I. Singh, PLA/banana fiber based sustainable bio-composites: A manufacturing perspective, *Compos. Part B Eng.* 180 (2020) 107535. https://doi.org/10.1016/j.compositesb.2019.107535

73. S. Chaitanya, I. Singh, J. Il Song, Recyclability analysis of PLA/Sisal fiber bio-composites, *Compos. Part B Eng.* 173 (2019) 106895. https://doi.org/10.1016/j.compositesb.2019.05.106

74. E. C. Ramires, E. Frollini, Tannin-phenolic resins: Synthesis, characterization, and application as matrix in biobased composites reinforced with sisal fibers, *Compos. Part B Eng.* 43 (2012) 2851–2860. https://doi.org/10.1016/j.compositesb.2012.04.049

75. R. Ortega, M. D. Monzón, Z. C. Ortega, E. Cunningham, Study and fire test of banana fibre reinforced composites with flame retardance properties, *Open Chem.* 18 (2020) 275–286. https://doi.org/10.1515/chem-2020-0025

76. A. Naughton, M. Fan, J. Bregulla, Fire resistance characterisation of hemp fibre reinforced polyester composites for use in the construction industry, *Compos. Part B Eng.* 60 (2014) 546–554. https://doi.org/10.1016/j.compositesb.2013.12.014

77. M. Y. Khalid, A. Al Rashid, Z. U. Arif, W. Ahmed, H. Arshad, A. A. Zaidi, Natural fiber reinforced composites: Sustainable materials for emerging applications, *Results Eng.* 11 (2021) 100263. https://doi.org/10.1016/j.rineng.2021.100263

78. S. Gul, M. Awais, S. Jabeen, M. Farooq, Recent trends in preparation and applications of biodegradable polymer composites, *J. Renew. Mater.* 8 (2020) 1305–1326. https://doi.org/10.32604/jrm.2020.010037

79. M. Karamanlioglu, G. D. Robson, The influence of biotic and abiotic factors on the rate of degradation of poly(lactic) acid (PLA) coupons buried in compost and soil, *Polym. Degrad. Stab.* 98 (2013) 2063–2071. https://doi.org/10.1016/j.polymdegradstab.2013.07.004

80. N. Altaee, G. A. El-Hiti, A. Fahdil, K. Sudesh, E. Yousif, Biodegradation of different formulations of polyhydroxybutyrate films in soil, *Springerplus.* 5 (2016) 762. https://doi.org/10.1186/s40064-016-2480-2

81. A. N. Boyandin, S. V. Prudnikova, M. L. Filipenko, E. A. Khrapov, A. D. Vasil'ev, T. G. Volova, Biodegradation of polyhydroxyalkanoates by soil microbial communities of different structures and detection of PHA degrading microorganisms, *Appl. Biochem. Microbiol.* 48 (2012) 28–36. https://doi.org/10.1134/S0003683812010024

82. C.-S. Wu, F.-S. Yen, C.-Y. Wang, Polyester/natural fiber biocomposites: Preparation, characterization, and biodegradability, *Polym. Bull.* 67 (2011) 1605–1619. https://doi.org/10.1007/s00289-011-0509-9

83. S. Chaudhuri, R. Chakraborty, P. Bhattacharya, Optimization of biodegradation of natural fiber (Chorchorus capsularis): HDPE composite using response surface methodology, *Iran. Polym. J.* 22 (2013) 865–875. https://doi.org/10.1007/s13726-013-0185-8

84. T. Fakhrul, M. A. Islam, Degradation behavior of natural fiber reinforced polymer matrix composites, *Procedia Eng.* 56 (2013) 795–800. https://doi.org/10.1016/j.proeng.2013.03.198

85. B. H. Briese, D. Jendrossek, H. G. Schlegel, Degradation of poly(3-hydroxybu-tyrate-co-3-hydroxyvalerate) by aerobic sewage sludge, *FEMS Microbiol. Lett.* 117 (1994) 107–111. https://doi.org/10.1111/j.1574-6968.1994.tb06750.x

86. V. T. Breslin, Degradation of starch-plastic composites in a municipal solid waste landfill, *J. Environ. Polym. Degrad.* 1 (1993) 127–141. https://doi.org/10.1007/BF01418206

87. E. Bari, J. J. Morrell, A. Sistani, Durability of natural/synthetic/biomass fiber-based polymeric composites, in: *Durab. Life Predict. Biocomposites, Fibre-Reinforced Compos. Hybrid Compos.*, Elsevier, 2019: pp. 15–26. https://doi.org/10.1016/B978-0-08-102290-0.00002-7

88. B. Aaliya, K. V. Sunooj, M. Lackner, Biopolymer composites: A review, *Int. J. Biobased Plast.* 3 (2021) 40–84. https://doi.org/10.1080/24759651.2021.1881214

89. M. Elkington, D. Bloom, C. Ward, A. Chatzimichali, K. Potter, Hand layup: Understanding the manual process, *Adv. Manuf. Polym. Compos. Sci.* 1 (2015) 138–151. https://doi.org/10.1080/20550340.2015.1114801

90. M. Raji, H. Abdellaoui, H. Essabir, C.-A. Kakou, R. Bouhfid, A. el kacem Qaiss, Prediction of the cyclic durability of woven-hybrid composites, in: *Durab. Life Predict. Biocomposites, Fibre-Reinforced Compos. Hybrid Compos.*, Elsevier, 2019, pp. 27–62. https://doi.org/10.1016/B978-0-08-102290-0.00003-9

91. M. Yuhazri, H. Sihombing, A comparison process between vacuum infusion and hand lay-up method toward kenaf/polyester composite, *Int. J. Basic Appl. Sci.* 10 (2010) 63–66.

92. G. D. Shrigandhi, B. S. Kothavale, Biodegradable composites for filament winding process, *Mater. Today Proc.* 42 (2021) 2762–2768. https://doi.org/10.1016/j.matpr.2020.12.718

93. P. Khalili, K. Y. Tshai, D. Hui, I. Kong, Synergistic of ammonium polyphosphate and alumina trihydrate as fire retardants for natural fiber reinforced epoxy composite, *Compos. Part B Eng.* 114 (2017) 101–110. https://doi.org/10.1016/j.compositesb.2017.01.049

94. A. M. Fairuz, S. M. Sapuan, E. S. Zainudin, C. N. A. Jaafar, Pultrusion process of natural fibre-reinforced polymer composites, in: M. S. Salit, M. Jawaid, N. Bin Yusoff, M. E. Hoque (Eds.), *Manuf. Nat. Fibre Reinf. Polym. Compos.*, Springer International Publishing, Cham, 2015: pp. 217–231. https://doi.org/10.1007/978-3-319-07944-8_11

95. Y. W. Leong, S. Thitithanasarn, K. Yamada, H. Hamada, Compression and injection molding techniques for natural fiber composites, in: *Nat. Fibre Compos.*, Elsevier, 2014: pp. 216–232. https://doi.org/10.1533/9780857099228.2.216

96. E. G. Kim, J. K. Park, S. H. Jo, A study on fiber orientation during the injection molding of fiber-reinforced polymeric composites, *J. Mater. Process. Technol.* 111 (2001) 225–232. https://doi.org/10.1016/S0924-0136(01)00521-0

97. M. Carus, C. Gahle, Injection moulding with natural fibres, *Reinf. Plast.* 52 (2008) 18–25. https://doi.org/10.1016/S0034-3617(08)70101-2

98. K. L. Pickering, M. G. A. Efendy, T. M. Le, A review of recent developments in natural fibre composites and their mechanical performance, *Compos. Part A Appl. Sci. Manuf.* 83 (2016) 98–112. https://doi.org/10.1016/j.compositesa.2015.08.038

99. D. Abraham, S. Matthews, R. McIlhagger, A comparison of physical properties of glass fibre epoxy composites produced by wet lay-up with autoclave consolidation and resin transfer moulding, *Compos. Part A Appl. Sci. Manuf.* 29 (1998) 795–801. https://doi.org/10.1016/S1359-835X(98)00055-4

100. N. Kumari, M. Paswan, K. Prasad, Effect of sawdust addition on the mechanical and water absorption properties of banana-sisal/epoxy natural fiber composites, *Mater. Today Proc.* (2021). https://doi.org/10.1016/j.matpr.2021.07.489

101. C. M. Meenakshi, A. Krishnamoorthy, Preparation and mechanical characterization of flax and glass fiber reinforced polyester hybrid composite laminate by hand lay-up method, *Mater. Today Proc.* 5 (2018) 26934–26940. https://doi.org/10.1016/j.matpr.2018.08.181

102. L. Yan, N. Chouw, K. Jayaraman, Lateral crushing of empty and polyurethane-foam filled natural flax fabric reinforced epoxy composite tubes, *Compos. Part B Eng.* 63 (2014) 15–26. https://doi.org/10.1016/j.compositesb.2014.03.013

103. B. Middleton, Composites: Manufacture and Application, in: *Des. Manuf. Plast. Components Multifunct.*, Elsevier, 2016: pp. 53–101. https://doi.org/10.1016/B978-0-323-34061-8.00003-X

104. J. Holbery, D. Houston, Natural-fiber-reinforced polymer composites in automotive applications, *JOM.* 58 (2006) 80–86. https://doi.org/10.1007/s11837-006-0234-2

105. M. K. Gupta, R. K. Srivastava, Mechanical properties of hybrid fibers-reinforced polymer composite: A review, *Polym. Plast. Technol. Eng.* 55 (2016) 626–642. https://doi.org/10.1080/03602559.2015.1098694

106. I. Mustapa, R. Shanks, I. Kong, Mechanical properties of self-reinforced poly(lactic acid) composites prepared by compression moulding of non-woven precursors, *World J. Eng.* 10 (2013) 23–28. https://doi.org/10.1260/1708-5284.10.1.23

107. P. Joseph, Effect of processing variables on the mechanical properties of sisal-fiber-reinforced polypropylene composites, *Compos. Sci. Technol.* 59 (1999) 1625–1640. https://doi.org/10.1016/S0266-3538(99)00024-X

108. R. Siva, S. Sundar Reddy Nemali, S. Kishore Kunchapu, K. Gokul, T. Arun Kumar, Comparison of mechanical properties and water absorption test on injection molding and extrusion – Injection molding thermoplastic hemp fiber composite, *Mater. Today Proc.* (2021). https://doi.org/10.1016/j.matpr.2021.05.189

109. H. Ardebili, J. Zhang, M. G. Pecht, Injection molding, in: *Encapsulation Technol. Electron. Appl.*, Elsevier, 2019: pp. 183–194. https://doi.org/10.1016/B978-0-12-811978-5.00004-3

110. R. R. P. Kuppusamy, S. Rout, K. Kumar, Advanced manufacturing techniques for composite structures used in aerospace industries, in: *Mod. Manuf. Process.*, Elsevier, 2020: pp. 3–12. https://doi.org/10.1016/B978-0-12-819496-6.00001-4

111. G. Francucci, E. S. Rodríguez, A. Vázquez, Experimental study of the compaction response of jute fabrics in liquid composite molding processes, *J. Compos. Mater.* 46 (2012) 155–167. https://doi.org/10.1177/0021998311410484

112. M. Shah, V. Chaudhary, Flow modeling and simulation study of vacuum assisted resin transfer molding (VARTM) process: A review, *IOP Conf. Ser. Mater. Sci. Eng.* 872 (2020) 012087. https://doi.org/10.1088/1757-899X/872/1/012087

113. M. R. Ricciardi, V. Antonucci, M. Durante, M. Giordano, L. Nele, G. Starace, A. Langella, A new cost-saving vacuum infusion process for fiber-reinforced composites: Pulsed infusion, *J. Compos. Mater.* 48 (2014) 1365–1373. https://doi.org/10.1177/0021998313485998

114. K. Han, S. Jiang, C. Zhang, B. Wang, Flow modeling and simulation of SCRIMP for composites manufacturing, *Compos. Part A Appl. Sci. Manuf.* 31 (2000) 79–86. https://doi.org/10.1016/S1359-835X(99)00053-6

115. M. Quanjin, M. R. M. Rejab, J. Kaige, M. S. Idris, M. N. Harith, Filament winding technique, experiment and simulation analysis on tubular structure, *IOP Conf. Ser. Mater. Sci. Eng.* 342 (2018) 012029. https://doi.org/10.1088/1757-899X/342/1/012029

116. Q. Wang, T. Li, B. Wang, C. Liu, Q. Huang, M. Ren, Prediction of void growth and fiber volume fraction based on filament winding process mechanics, *Compos. Struct.* 246 (2020) 112432. https://doi.org/10.1016/j.compstruct.2020.112432

117. A. Memon, A. Nakai, The processing design of jute spun yarn/PLA braided composite by pultrusion molding, *Adv. Mech. Eng.* 5 (2013) 816513. https://doi.org/10.1155/2013/816513

118. A. M. Fairuz, S. M. Sapuan, E. S. Zainudin, C. N. A. Jaafar, Optimization of Pultrusion Process for Kenaf Fibre Reinforced Vinyl Ester Composites, in: *Recent Technol. Des. Manag. Manuf.*, Trans Tech Publications Ltd, 2015: pp. 499–503. https://doi.org/10.4028/www.scientific.net/AMM.761.499

119. I. Baran, C. C. Tutum, J. H. Hattel, The effect of thermal contact resistance on the thermosetting pultrusion process, *Compos. Part B Eng.* 45 (2013) 995–1000. https://doi.org/10.1016/j.compositesb.2012.09.049

120. P. V. Joseph, M. S. Rabello, L. H. Mattoso, K. Joseph, S. Thomas, Environmental effects on the degradation behaviour of sisal fibre reinforced polypropylene composites, *Compos. Sci. Technol.* 62 (2002) 1357–1372. https://doi.org/10.1016/S0266-3538(02)00080-5

121. A. Q. Dayo, A. A. Babar, Q. Qin, S. Kiran, J. Wang, A. H. Shah, A. Zegaoui, H. A. Ghouti, W. Liu, Effects of accelerated weathering on the mechanical properties of hemp fibre/polybenzoxazine based green composites, *Compos. Part A Appl. Sci. Manuf.* 128 (2020) 105653. https://doi.org/10.1016/j.compositesa.2019.105653

122. P. Fei, H. Xiong, J. Cai, C. Liu, Zia-ud-Din, Y. Yu, Enhanced the weatherability of bamboo fiber-based outdoor building decoration materials by rutile nano-TiO$_2$, *Constr. Build. Mater.* 114 (2016) 307–316. https://doi.org/10.1016/j.conbuildmat.2016.03.166

123. E. Omrani, P. L. Menezes, P. K. Rohatgi, State of the art on tribological behavior of polymer matrix composites reinforced with natural fibers in the green materials world, *Eng. Sci. Technol. an Int. J.* 19 (2016) 717–736. https://doi.org/10.1016/j.jestch.2015.10.007

124. M. Brebu, Environmental degradation of plastic composites with natural fillers—A review, *Polymers (Basel)*. 12 (2020) 166. https://doi.org/10.3390/polym12010166

125. N. P. J. Dissanayake, J. Summerscales, S. M. Grove, M. M. Singh, Life cycle impact assessment of flax fibre for the reinforcement of composites, *J. Biobased Mater. Bioenergy.* 3 (2009) 245–248. https://doi.org/10.1166/jbmb.2009.1029

126. M. George, D. C. Bressler, Comparative evaluation of the environmental impact of chemical methods used to enhance natural fibres for composite applications and glass fibre based composites, *J. Clean. Prod.* 149 (2017) 491–501. https://doi.org/10.1016/j.jclepro.2017.02.091

127. M. Li, Y. Pu, V. M. Thomas, C. G. Yoo, S. Ozcan, Y. Deng, K. Nelson, A. J. Ragauskas, Recent advancements of plant-based natural fiber-reinforced composites and their applications, *Compos. Part B Eng.* 200 (2020) 108254. https://doi.org/10.1016/j.compositesb.2020.108254

128. C. Badji, J. Beigbeder, H. Garay, A. Bergeret, J.-C. Bénézet, V. Desauziers, Correlation between artificial and natural weathering of hemp fibers reinforced polypropylene biocomposites, *Polym. Degrad. Stab.* 148 (2018) 117–131. https://doi.org/10.1016/j.polymdegradstab.2018.01.002

129. S. S. Chee, M. Jawaid, M. T. H. Sultan, O. Y. Alothman, L. C. Abdullah, Accelerated weathering and soil burial effects on colour, biodegradability and thermal properties of bamboo/kenaf/epoxy hybrid composites, *Polym. Test.* 79 (2019) 106054. https://doi.org/10.1016/j.polymertesting.2019.106054

130. M. Yu, Y. Zheng, J. Tian, Study on the biodegradability of modified starch/polylactic acid (PLA) composite materials, *RSC Adv.* 10 (2020) 26298–26307. https://doi.org/10.1039/D0RA00274G

131. M. Assarar, D. Scida, A. El Mahi, C. Poilâne, R. Ayad, Influence of water ageing on mechanical properties and damage events of two reinforced composite materials: Flax-fibres and glass-fibres, *Mater. Des.* 32 (2011) 788–795. https://doi.org/10.1016/j.matdes.2010.07.024

132. C. Chow, X. Xing, R. Li, Moisture absorption studies of sisal fibre reinforced polypropylene composites, *Compos. Sci. Technol.* 67 (2007) 306–313. https://doi.org/10.1016/j.compscitech.2006.08.005

133. H. T. Sreenivas, N. Krishnamurthy, G. R. Arpitha, A comprehensive review on light weight kenaf fiber for automobiles, *Int. J. Light. Mater. Manuf.* 3 (2020) 328–337. https://doi.org/10.1016/j.ijlmm.2020.05.003

134. O. Akampumuza, P. M. Wambua, A. Ahmed, W. Li, X.-H. Qin, Review of the applications of biocomposites in the automotive industry, *Polym. Compos.* 38 (2017) 2553–2569. https://doi.org/10.1002/pc.23847

135. N. Ayrilmis, S. Jarusombuti, V. Fueangvivat, P. Bauchongkol, R. H. White, Coir fiber reinforced polypropylene composite panel for automotive interior applications, *Fibers Polym.* 12 (2011) 919–926. https://doi.org/10.1007/s12221-011-0919-1

136. J. S. H. Deka, T. O. Varghese, S. K. Nayak, Recent development and future trends in coir fiber-reinforced green polymer composites: Review and evaluation, *Polym. Compos.* 37 (2016) 3296–3309. https://doi.org/10.1002/pc.23529

137. A. Gopinath, M. S. Kumar, A. Elayaperumal, Experimental investigations on mechanical properties of jute fiber reinforced composites with polyester and epoxy resin matrices, *Procedia Eng.* 97 (2014) 2052–2063. https://doi.org/10.1016/j.proeng.2014.12.448

138. K. Selvakumar, O. Meenakshisundaram, Mechanical and dynamic mechanical analysis of jute and human hair-reinforced polymer composites, *Polym. Compos.* 40 (2019) 1132–1141. https://doi.org/10.1002/pc.24818

139. G. Cicala, G. Recca, L. Oliveri, Y. Perikleous, F. Scarpa, C. Lira, A. Lorato, D. J. Grube, G. Ziegmann, Hexachiral truss-core with twisted hemp yarns: Out-of-plane shear properties, *Compos. Struct.* 94 (2012) 3556–3562. https://doi.org/10.1016/j.compstruct.2012.05.020

140. H. N. Dhakal, Z. Zhang, The use of hemp fibres as reinforcements in composites, in: *Biofiber Reinf. Compos. Mater.*, Elsevier, 2015: pp. 86–103. https://doi.org/10.1533/9781782421276.1.86

141. M. John, S. Thomas, Biofibres and biocomposites, *Carbohydr. Polym.* 71 (2008) 343–364. https://doi.org/10.1016/j.carbpol.2007.05.040
142. A. Muhammad, M. R. Rahman, R. Baini, M. K. Bin Bakri, Applications of sustainable polymer composites in automobile and aerospace industry, in: *Advances in Sustainable Polymer Composites.*, Elsevier, 2021: pp. 185–207. https://doi.org/10.1016/B978-0-12-820338-5.00008-4
143. M. R. Mansor, A. H. Nurfaizey, N. Tamaldin, M. N. A. Nordin, Natural fiber polymer composites, in: *Biomass, Biopolym. Mater. Bioenergy*, Elsevier, 2019: pp. 203–224. https://doi.org/10.1016/B978-0-08-102426-3.00011-4
144. U. P. Breuer, *Commercial Aircraft Composite Technology*, Springer International Publishing, Cham, 2016. https://doi.org/10.1007/978-3-319-31918-6
145. N. J. Arockiam, M. Jawaid, N. Saba, Sustainable bio composites for aircraft components, in: *Sustain. Compos. Aerosp. Appl.*, Elsevier, 2018: pp. 109–123. https://doi.org/10.1016/B978-0-08-102131-6.00006-2
146. S. Balck, Looking to lighten up aircraft interiors? Try natural fibers!, *Compos. World.* (2015). https://www.comositesworld.com/articles/looking-to-lighten-up-aircraft-interiors---with-natural-fibers
147. F. S. Eloy, R. R. Costa, R. De Medeiros, M. L. Ribeiro, V. Tita, Comparison between mechanical properties of bio and synthetic composites for use in aircraft interior structures, *Phys. Rev.*, 2015.
148. M. Fan, Future scope and intelligence of natural fibre based construction composites, in: *Adv. High Strength Nat. Fibre Compos. Constr.*, Elsevier, 2017: pp. 545–556. https://doi.org/10.1016/B978-0-08-100411-1.00022-4
149. V. Barbieri, M. Lassinantti Gualtieri, C. Siligardi, Wheat husk: A renewable resource for bio-based building materials, *Constr. Build. Mater.* 251 (2020) 118909. https://doi.org/10.1016/j.conbuildmat.2020.118909
150. S. N. Chinnu, S. N. Minnu, A. Bahurudeen, R. Senthilkumar, Reuse of industrial and agricultural by-products as pozzolan and aggregates in lightweight concrete, *Constr. Build. Mater.* 302 (2021) 124172. https://doi.org/10.1016/j.conbuildmat.2021.124172
151. V. S. Nadh, G. S. Vignan, K. Hemalatha, A. Rajani, Mechanical and durability properties of treated oil palm shell lightweight concrete, *Mater. Today Proc.* (2021). https://doi.org/10.1016/j.matpr.2021.04.373
152. A. Albers, J. Holoch, S. Revfi, M. Spadinger, Lightweight design in product development: A conceptual framework for continuous support in the development process, *Procedia CIRP.* 100 (2021) 494–499. https://doi.org/10.1016/j.procir.2021.05.109
153. K. N. Bharath, S. Basavarajappa, Applications of biocomposite materials based on natural fibers from renewable resources: A review, *Sci. Eng. Compos. Mater.* 23 (2016) 123–133. https://doi.org/10.1515/secm-2014-0088
154. C. Asasutjarit, J. Hirunlabh, J. Khedari, S. Charoenvai, B. Zeghmati, U. C. Shin, Development of coconut coir-based lightweight cement board, *Constr. Build. Mater.* 21 (2007) 277–288. https://doi.org/10.1016/j.conbuildmat.2005.08.028
155. K. Goda, M. S. Sreekala, A. Gomes, T. Kaji, J. Ohgi, Improvement of plant based natural fibers for toughening green composites—Effect of load application during mercerization of ramie fibers, *Compos. Part A Appl. Sci. Manuf.* 37 (2006) 2213–2220. https://doi.org/10.1016/j.compositesa.2005.12.014

156. A. M. Youssef, M. A. El-Samahy, M. H. Abdel Rehim, Preparation of conductive paper composites based on natural cellulosic fibers for packaging applications, *Carbohydr. Polym.* 89 (2012) 1027–1032. https://doi.org/10.1016/j.carbpol.2012.03.044

157. N. Uddin, R. R. Kalyankar, Manufacturing and structural feasibility of natural fiber reinforced polymeric structural insulated panels for panelized construction, *Int. J. Polym. Sci.*, 2011 (2011) 1–7. https://doi.org/10.1155/2011/963549

158. I. Sanal, D. Verma, Construction Materials Reinforced with Natural Products, in: *Handb. Ecomater.*, Springer International Publishing, Cham, 2018: pp. 1–24. https://doi.org/10.1007/978-3-319-48281-1_75-1

159. A. Amiri, T. Krosbakken, W. Schoen, D. Theisen, C. A. Ulven, Design and manufacturing of a hybrid flax/carbon fiber composite bicycle frame, *Proc. Inst. Mech. Eng. Part P J. Sport. Eng. Technol.* 232 (2018) 28–38. https://doi.org/10.1177/1754337117716237

160. H. Mason, Sicomin bio-based resins used in flax fiber-reinforced boat, *Compos. World.* (2020). https://www.compositesworld.com/news/sicomin-bio-based-resins-used-in-flax-fiber-reinforced-boat-

Chapter 9

Sustainable composites for lightweight applications

A. Soundhar
Indian Institute of Technology Guwahati, Guwahati, India

V. Lakshmi Narayanan
Vellore Institute of Technology, Vellore, India

M. Natesh
V.S.B. Engineering College, Karur, India

K. Jayakrishna
Vellore Institute of Technology, Vellore, India

CONTENTS

DOI: 10.1201/9781003252108-9

9.1 INTRODUCTION TO SUSTAINABLE COMPOSITES

A composite is a material produced from two or more different classes of materials combined to offer excellent properties (physical, chemical, and mechanical), depending upon the specific application [1]. In engineering applications, composites are used because of their high performance with respect to enhanced specific stiffness and strength (strength-to-weight ratio), which aids in lessening the overall weight of the parts. For example, lessening overall weight leads to a reduction in fuel consumption in automobiles, while also providing improved performance and decreased CO_2 emissions [2]. The performance of composite materials is affected by the different parameters.

Based on the volume of the composite, reinforcement type, processing method, and geometry, significant properties (fracture toughness behavior [mode I, mode II, and mixed-mode], impact resistance, tensile strength and modulus, vibration behavior related to damping, coefficient of thermal expansion [CTE], thermal properties such as thermal decomposition, and thermal conductivity) are obtained [3, 4]. Hence, realizing these key factors is significant in the process of design, manufacturing, and optimum life of composites. While much research has been conducted on composites, their structure to property relationship is not completely understood based on volume percentage, reinforcement types, environmental aspects, and life of composites [5].

This chapter offers a brief introduction about fibers or reinforcements and their properties based on environmental impact. Four classes of composites (ceramic-matrix composites [CMCs], metal-matrix composites [MMCs], carbon matrix composites, and fiber-reinforced polymer matrix composites [FRPCs]) are available. In these composites, metal and ceramic matrix composites are used in high compression loading-related applications at elevated temperatures and do not experience tensile or impact loads [6].

Glass fiber-reinforced composites and carbon matrix composites are utilized in numerous industrial applications due to their versatile properties, high performance, and because they are commercially available [7]. They are also employed in the automotive, aerospace, and construction industries, as well as marine and sports equipment. In different classes of composites, fiber-reinforced polymers or polymer matrix composites are generally used in various applications [8, 9]. PMCs are utilized with both natural and synthetic fibers. Natural fibers such as jute, flax, hemp, date palm kenaf, and many others are used as reinforcement materials. Carbon, aramids, and glass fibers are known as synthetic fibers [10].

Various industries (structural and non-structural applications) utilize composites without using metallic materials due to their advantages (high performance and excellent properties) [11]. Today, environmentally friendly materials are expected to be used to help make the world a more hazard-free place. Thus, as glass and carbon fibers are non-biodegradable, alternative biodegradable fibers for composites are needed [12]. Hence, researchers have focused on natural fiber-reinforced composites because of their significant properties (high specific tensile strength and modulus) and user-friendly benefits such as being biodegradable, inexpensive, lightweight, and easily available.

Life-cycle assessment (LCA) illustrates the environmental impact among glass, carbon, and natural fibers in composite production [13, 14]. One characteristic of natural fiber-reinforced composites is water absorption (hydrophilic), which reduces the mechanical properties of the composites that limit their use in several structural and non-structural applications [15, 16]. Certain chemical treatments can be applied to natural fibers that prevent them from absorbing water (hydrophobic). These surface treatments make natural fibers compatible with thermoplastic and thermoset polymer matrices, which improves the physical, mechanical, and thermal properties of the composites [17].

This chapter will discuss natural fiber reinforcements and their morphology, design and manufacturing techniques of biocomposites, sustainable methods for improving the properties, and future scope of biocomposites based on sustainability.

9.2 SUSTAINABLE NATURAL FIBER REINFORCEMENTS AND THEIR MORPHOLOGICAL STRUCTURES

Lightweight composite materials reinforced with plant fibers have been explored as an alternative to traditional fiber-reinforcing materials due to environmental concerns, regulations for low-emission products, and consumer awareness for sustainable products [18]. Both matrices and reinforcements are expected to be derived from renewable resources in order to make composite materials more sustainable and eco-friendly. Natural fibers for composite reinforcement are divided into three main types: lignocellulose plant-based (hemp, flax, jute sisal, palm fiber, kenaf, date palm, etc.), animal-based (silk, wool, and hair), and mineral-derived (asbestos, wollastonite) [19].

Despite their many appealing characteristics, natural plant fibers have not been utilized to their full potential. Due to their intrinsic moisture absorption behavior (cellulose, hemicelluloses, lignin, and pectin) and natural variability, natural plant fiber-reinforced composites have been utilized mostly for non-structural applications in the automotive, construction, and packaging sectors [20]. The moisture absorption behavior of these fibers results in poor fiber matrix adherence, causing non-uniform stress distribution from matrix to fibers under different loading conditions [21]. Many researchers have conducted different fiber surface modification procedures, along with better manufacturing processes, to address such issues [22–24]. The most commonly used natural fibers (along with some that are plant-based), as well as their major features, will be discussed in the following sub-sections.

9.2.1 Hemp fiber

Hemp (Cannabis Sativa L.) is a natural fiber that is among the oldest, strongest, and stiffest natural fibers. The chemical components of mature hemp fiber include cellulose (74%), hemicellulose (18%), lignin (4%), and pectin (1%). Hemp has strong environmental and performance appeal as a reinforcing element in composites, particularly for automotive applications. However, when compared to glass fiber-reinforced plastic composites, its main drawback is poor impact strength [25]. Figure 9.1 shows a non-woven hemp mat and a scanning electron microscopy (SEM) image of hemp fiber.

9.2.2 Sisal fiber

One of the most commonly employed natural fibers in the application of yarns, ropes, twines, belts, rugs, carpets, mattresses, mats, and handcrafted pieces is sisal fiber (agave sisalana), as shown in Figure 9.2. Until flowering, a sisal plant yields between 200 and 250 leaves, each of which includes about

Figure 9.1 Non-woven hemp mat and SEM image of hemp fiber.

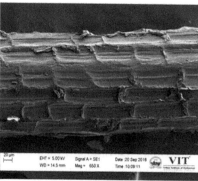

Figure 9.2 Woven sisal fiber mat and the SEM image of sisal fiber.

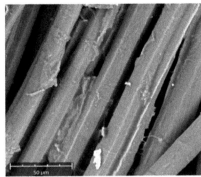

Figure 9.3 Woven flax fiber mat and the SEM image of flax fiber.

700–1400 fiber bundles with a length of about 0.5–1.0 m [26] (Soundhar and Jayakrishna 2021). The sisal leaf consists of 4% fiber, 1% cuticle, 8% dry matter, and 87% water in a sandwich form. Sisal is environmentally friendly, as it is biodegradable. World production of sisal fiber is about 300,000 tons per year.

9.2.3 Flax fiber

Flax (Linum usitatissimum L.) is an annual plant that grows to a height of 0.5–1.5 m. The cellulose content in flax fibers is approximately 71%, hemicellulose 20.6%, lignin 2.2%, and 1.2% wax. Due to their high molecular weight and crystallinity of cellulose, flax fibers have greater mechanical characteristics such as strength and stiffness than other natural plant fibers (Figure 9.3). This fiber has been utilized in a variety of applications, including the automotive, marine, and textile sectors [27].

9.3 DESIGN AND MANUFACTURING PROCESS OF BIOCOMPOSITES

Different designs and manufacturing techniques have been used to improve the inherent characteristics of biocomposite materials. The selection of natural fibers or filler (reinforcements) and matrices used, as well as the consideration of sizes/part design and processing (fiber surface treatments, stacking sequences, fiber orientation, compaction methods, and curing techniques), are all part of these design and manufacturing processes [28, 29]. All these processes influence the mechanical, physical, electrical, and thermal characteristics of a specific biocomposite. As a result, improvement of biocomposite materials begins with the design stage and continues through the production phase, with all steps closely monitored to avoid defects in the prepared biocomposites [30].

Today's biocomposite design and production methods face several challenges. These include, but are not limited to, increased use of biocomposites, unpredicted material behaviors (mostly non-linear characteristics), and limited design data due to the various types or species of natural fibers and matrices [31]. Furthermore, interactions between hydrophilic natural fibers and hydrophobic matrices are critical problems affecting the design and manufacturing processes of numerous biocomposites. Hence, various design techniques and manufacturing methods will be considered in the manufacture of biocomposites.

9.3.1 Eco-design and sustainability

Products made from composite materials have several advantages, one of which is environmental friendliness owing to their sustainable sources, biodegradability, renewability, and corrosion resistance, as characterized by natural or biofibers [32]. Natural fiber-reinforced polymer composites (NFRPCs) have gained popularity in recent years as a low-cost, ecologically acceptable alternative to more widely used reinforcing materials such as glass and carbon fibers [33]. One of the most important elements influencing a material's sustainability is its cost. Several natural fibers—particularly jute, coir, and bamboo—are significantly less expensive (in terms of unit price) than several frequently used synthetic fibers like carbon and E-glass [34]. As a result, the total cost of producing natural FRP composites is reduced.

Sustainability is one of the major developmental concepts of the future generation of products (materials) and processes (manufacturing). As a viable alternative for synthetic, non-renewable fiber-reinforced polymer composites, this resulted in the development of sustainable, eco-efficient, biodegradable, and environmentally friendly biocomposite materials [35]. Land usage, energy efficiency, resource availability, soil conversation, biodiversity, environmental effect, and social community impact are all aspects of sustainability [27]. Natural fibers are abundant, sustainable, and biodegradable as a primary constituent/reinforcement of biocomposites. Hemp, jute, sisal, bamboo, kenaf, rice husk and straw, wheat straw, oil palm, henequen, curaua, olive pit/husk, coir, pineapple leaf, ramie, choir, abaca, flax, date palm, banana, pineapple, and bagasse are examples of natural fibers [36, 37]. For example, processing natural (hemp) fibers has a relatively low environmental effect, particularly in comparison with synthetic fibers (glass fiber). A comparison of hemp fiber with glass fiber is illustrated in Table 9.1.

9.3.2 Design for environment

Design for the environment is a method of accomplishing design goals without compromising or adversely affecting the environment or human health throughout the process or product life cycle [39]. It focuses on improving

Table 9.1 Comparison of environmental impact during the production of glass and hemp fibers (1 kg) [38]

Parameters	Fibers	
	Hemp	Glass
Power consumption (MJ)	3.5	48.5
NO_x emission (g)	0.96	2.8
SO_x emission (g)	1.1	8.7
CO_2 emission (kg)	0.65	20.5
BOD (mg)	0.266	1.76

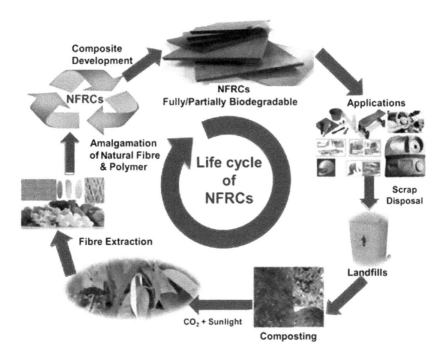

Figure 9.4 Product life cycle stages.

product quality and affordability while reducing or eradicating all environmental consequences of a product throughout its life cycle. Biocomposites involve five different product life-cycle stages such as materials, manufacturing, distribution, usage, and recovery, as illustrated in Figure 9.4. The processes involved in each stage are appropriately planned and monitored to maintain or safeguard a healthy environment [40, 41].

9.3.3 Design for manufacture

The potential for biocomposite materials to be used in a wide range of engineering applications has highlighted the importance of design for manufacture. Furthermore, biocomposite design for manufacture is expected to promote cost reductions in manufacturing, components, and assembly units [42]. There are numerous methods for producing various biocomposite materials on both a laboratory/small and commercial/large scale such as hand and spray lay-ups, injection molding, compression molding, extrusion, resin transfer molding, vacuum bagging, filament winding, automated fiber placement, autoclave and out-of-autoclave, as well as additive manufacturing [43, 44].

9.4 SUSTAINABILITY TECHNIQUES FOR PROPERTY ENHANCEMENT

With recent government laws recommending, promoting, and encouraging the use of renewable energy to prevent pollution of the environment, many composite manufacturing industries will switch to the manufacture of bio-composites [45]. This is due to the fact that producing biocomposites uses less energy and emits less CO_2 and other harmful gases into the atmosphere, as compared to the production of traditional composites from synthetic carbon and glass fibers, as well as petroleum-based matrices [46].

When it comes to the health of the employees, the biocomposite process is almost risk-free. An environmental life-cycle study of a whole product (composite) includes five key stages: raw material extraction, manufacture, distribution, usage, and disposal/recovery. The first four steps have been directly and indirectly elucidated [47]. Now, considering the last stage of product disposal/recovery after usage, biocomposite materials' partial and full biodegradability provides a viable disposal method. Because of their composition, biocomposites are readily decomposed and return to the soil [48].

The chemical and structural compositions of numerous natural fibers have a major impact on their inherent properties. These two elements are the functions of climatic conditions, harvest time, extraction process, fiber treatment, and storage techniques [49].

9.4.1 Improvement of reinforcements and matrices through various treatments and fillers

Due to some limitations of natural fiber-reinforced polymer (FRP) biocomposite materials, it is required to treat natural fibers so as to overcome the problems of weak interfacial adhesion established with polymers, hydrophilicity, and low thermal stability [50]. Much research has been conducted to enhance the characteristics of natural fibers and, subsequently, improve compatibility between fibers and matrices [51, 52]. One of the primary reasons for various natural fiber surface treatments is to improve the wettability of the fibers so that matrices may adequately wet the fibers, resulting in improved adhesion between the fibers and matrices. Several methods (chemical, physical, additive, and biological treatments) have been employed to improve the interfacial bonding between fiber and matrix interface [53].

9.4.1.1 Chemical treatments

It is widely understood that the surface of hydrophilic cellulose fibers and hydrophobic polymeric matrix are incompatible. Importantly, the chemical modification technique aids in the compatibility of fibers and matrix, resulting in a considerable increase in wetting efficiency [54]. Therefore, the surfaces of natural fibers are chemically treated to increase the fiber-polymer interfacial bonding or adhesion by minimizing their incompatibility. Alkali, zirconate, acryl, peroxide, titanate, isocyanate, permanganate, acrylonitrile, benzyl, and acetyl treatments, as well as maleated anhydride grafted coupling agents, are used to chemically modify the natural fibers [55]. The most commonly used method is alkaline treatment using sodium hydroxide (NaOH) followed by silane, acetyl, and maleated anhydride grafted coupling agents [56]. The alkaline treatment eliminates not only hemicellulose and lignin, but also waxes, fatty acids, and other organic residues from the fiber surface.

9.4.1.2 Physical treatments

Physical treatment techniques are mostly used to enhance the fiber-matrix adhesion by increasing surface properties of the fibers. Physical treatments such as plasma, corona discharge, ultraviolet (UV), fiber beating, heat treatment, and electron radiation can improve the functional surface and structural characteristics of natural fibers [54]. To improve the surface roughness of natural fibers, plasma treatments can be performed at a low temperature, which increases the surface roughness of the fiber by increasing hydrophobicity at the fiber surface. As a result, interfacial adhesion improves.

The corona treatment has also been used to greatly enhance the wettability of some natural fibers, such as woods and the surface energy of cellulosic fibers [43–45]. It produces physical and chemical changes in the fibers, such as increased surface polarity and fiber roughness. Neto et al. (2021) [57] compared the corona and UV treatments on natural fibers to enhance the mechanical properties of jute fibers/epoxy biocomposite (flexural and tensile strengths, storage modulus, polarity, and tenacity). The polarity of the fibers was enhanced as a result of these treatments.

The corona treatment implies reduced flexural strength of the jute fibers, whereas UV treatment enhanced the biocomposite's flexural strength by approximately 30% under ideal treatment circumstances [52]. The fiber beating technique causes defibrillation of the fibers, which increases their surface area, allowing for better mechanical interlocking. Hydrothermal treatments can also reduce both hydrophilicity (rate of moisture uptake) and thermal degradation of natural fibers [55].

9.4.2 Hybridization

Hybridization offers a significant improvement in the properties of composites, which attracts attention from researchers for structural applications. In general, a significant weight percentage of synthetic fibers such as glass, carbon, etc., hybridized to enhance the inherent properties of natural fiber composites for intended applications [55]. For compressive applications, jute fibers are reinforced with glass fiber; however, the hybridization effect is not significant for tensile load applications. Thus, it reveals that the procedure requires for applications regarding the hybridization of natural fiber with synthetic fiber to achieve the desired goals/applications.

The reason behind enhancement in hybridization performance increases the load carry behavior of composites compared to when they are reinforced in the matrix alone [56]. However, the addition of synthetic fiber improves the properties of natural fiber composite, and compatibility between fiber/filler and matrix influences the failure of the hybrid composites. The hybridizing material (sisal fiber and hemp fiber) can be assembled in different stacking sequences, as shown in Figure 9.5.

Through experimental studies, most researchers reported that strength enhancement depends on fiber loading, treatment, aspect ratio, and contact area [57, 58]. Due to hybrid composites' enhanced strength and land-bearing capacity, they are used in many automotive applications such as car interior parts and door panels. Some reported research on natural fiber hybrid polymer composites is illustrated in Table 9.2.

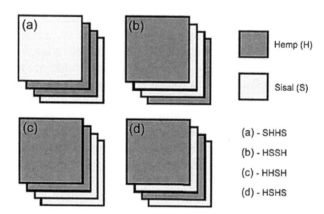

Figure 9.5 Different stacking sequences.

Table 9.2 Some reported research on natural fiber hybrid polymer composites

	Fiber	Matrix	Reference
Natural fiber and natural fiber	Jute/hemp/flax	Epoxy	Chaudhary et al. 2018 [59]
	Banana/coconut fiber	Polyester	Senthil Kumar et al. 2016 [60]
	Eucalyptus fiber/areca	Epoxy	Manjunath and Udupa 2016 [61]
	Jute/ramie	Epoxy	Mohanavel et al. 2021 [62]
	Sugarcane Bagasse/bamboo	Polyurethane foam	Abedom et al. 2021 [63]
	Jute/aloe vera	Epoxy	Chandramohan et al. 2020 [64]
	Caryota/sisal	Epoxy	Atmakuri et al. 2021 [65]
	Ramie/sisal/curaua	Epoxy	Pereira et al. 2020 [66]
	Kenaf/pineapple leaf fiber	Phenolic resin	Asim et al. 2019 [67]
	Sisal/hemp	Poly (lactic acid)	Pappu et al. 2019 [68]
Natural fiber and synthetic fiber	Sisal fiber/glass	Epoxy	Soundhar et al. 2020 [69]
	Sugar palm fibers/carbon	Epoxy	Baihaqi et al. 2021 [70]
	Kenaf fiber /kevlar fiber	Epoxy	Ramasamy et al. 2021 [71]
	Flax fiber/glass	Phenolic	Hajiha and Sain 2015 [72]
	Kenaf fiber /aramid	Epoxy	Yahaya et al. 2015 [73]
	Areca sheath/jute/glass	Epoxy	Jothibasu et al. 2018 [74]
	Basalt/glass fiber	Unsaturated polyester	Sapuan et al. 2020 [75]

9.4.2.1 Intra-ply hybridization

Researchers have investigated the hybridization of natural fiber with synthetic fiber and cellulosic fiber with natural fiber in a polymer matrix and found that hybridization improves the mechanical characteristics of natural fiber composites when compared to natural fiber composites reinforced only with natural fibers in the matrix [76]. Inter- and intra-ply hybridization is an approach to enhance the properties of the natural fiber composite. In hybridization, two or more natural fiber yarns are oriented in the warp and weft direction, which avoids the stacking sequence and fiber orientation of natural fiber in the matrix [77]. Figure 9.6 shows the different types of intra-ply hybrid woven mat (natural-natural, natural-glass) used in the composite field.

9.5 FUTURE OUTLOOKS AND CHALLENGES

9.5.1 Journey of composite materials towards sustainability

One of the primary aspects of sustainable lightweight composites is to reduce component weight, with materials employed expected to deliver both weight savings and required functional characteristics. This is a

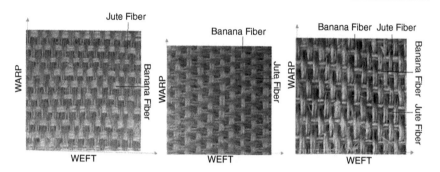

Figure 9.6 Different types of intra-ply hybrid woven mat.

revolutionary new technique that enables effective materials utilization—a high strength-to-weight ratio in contrast to metal counterparts—which leads to higher fuel consumption and environmental advantages by considerably reducing the carbon footprint. The use of advanced lightweight composite materials derived from renewable sources has increased significantly in recent decades, particularly in high-tech industry sectors such as automotive, aerospace, marine, and construction in the quest to reduce dependence on heavy metal parts and non-renewable conventional composites in favor of more sustainable composite materials with a lower carbon footprint. When compared to glass and carbon fiber-reinforced composites, sustainable lightweight composites have a significantly less environmental effect in terms of global warming potential (GWP), ecotoxicity, natural resource depletion, and acidification.

9.5.2 Market outlook and supply chain scenario

Composites have a wide range of applications due to their desirable characteristics such as high strength-to-weight ratio, excellent chemical and corrosion resistance, and compliance with "light weighting" materials. The development of lightweight composites has prompted high-tech industries such as automotive, marine, aerospace, sports leisure, construction, and wind energy to consider how they might profit from this new class of lightweight, environmentally friendly materials. In today's economy, composite materials have a large market share, and the automobile industry is one of the most important users of lightweight composites. Carbon fiber and natural fiber (hemp, sisal, kenaf, and ramie) reinforcements are used by major original equipment manufacturers (OEMs) such as Toyota, Mercedes-Benz, BMW, Audi, Ford, Hyundai, and Tata Motors, among others, to reduce the total mass of vehicles, resulting in lower CO_2 emissions during a product's lifecycle.

9.5.3 Future outlook

The potential for lightweight and ultra-lightweight composite materials to be used and developed in the future is an exciting opportunity for addressing sustainable materials and a low-carbon economy (LCE). Understanding design, material selection, and cost-effective manufacturing processes—acceptable strength and modulus, failure causes, and material structure-property is very important. Some advanced materials, manufacturing, and characterization methods are emerging to accomplish these goals. In this scenario, the paradigm changes from a fossil-fuel-based economy to a circular economy—and interdisciplinary research initiatives incorporating key stakeholders such as legislative bodies, OEMs, academic institutions, and consumers must be encouraged.

REFERENCES

1. Hodzic, Alma, and Robert Shanks, eds. *Natural Fibre Composites: Materials, Processes and Properties*. Woodhead Publishing, 2014.
2. Mallick, P. K. "Advanced materials for automotive applications: An overview," *Advanced Materials in Automotive Engineering* 1 (2012): 5–27.
3. Friedrich, Klaus, and Abdulhakim A. Almajid. "Manufacturing aspects of advanced polymer composites for automotive applications," *Applied Composite Materials* 20, no. 2 (2013): 107–128.
4. Ravishankar, B., Sanjay K. Nayak, and M. Abdul Kader. "Hybrid composites for automotive applications–A review," *Journal of Reinforced Plastics and Composites* 38, no. 18 (2019): 835–845.
5. Fan, J., and J. Njuguna. "An introduction to lightweight composite materials and their use in transport structures," In *Lightweight Composite Structures in Transport*, pp. 3–34. Woodhead Publishing, 2016.
6. Gorbatikh, Larissa, Brian L. Wardle, and Stepan V. Lomov. "Hierarchical lightweight composite materials for structural applications," *Mrs Bulletin* 41, no. 9 (2016): 672–677.
7. Agarwal, Jyoti, Swarnalata Sahoo, Smita Mohanty, and Sanjay K. Nayak. "Progress of novel techniques for lightweight automobile applications through innovative eco-friendly composite materials: a review," *Journal of Thermoplastic Composite Materials* 33, no. 7 (2020): 978–1013.
8. Patel, Murlidhar, Bhupendra Pardhi, Sulabh Chopara, and Manoj Pal. "Lightweight composite materials for automotive – a review," *Carbon* 1, no. 2500 (2018): 151.
9. Thakur, Vijay Kumar, Manju Kumari Thakur, Prasanth Raghavan, and Michael R. Kessler. "Progress in green polymer composites from lignin for multifunctional applications: A review," *ACS Sustainable Chemistry & Engineering* 2, no. 5 (2014): 1072–1092.
10. Sarikaya, Engin, Hasan Çalioğlu, and Hakan Demirel. "Production of epoxy composites reinforced by different natural fibers and their mechanical properties," *Composites Part B: Engineering* 167 (2019): 461–466.

11. Kumar, Rajiv, Mir Irfan Ul Haq, Ankush Raina, and Ankush Anand. "Industrial applications of natural fibre-reinforced polymer composites–challenges and opportunities," *International Journal of Sustainable Engineering* 12, no. 3 (2019): 212–220.
12. Kumar, Rajiv, and Ankush Anand. "Fabrication and mechanical characterization of Indian ramie reinforced polymer composites," *Materials Research Express* 6, no. 5 (2019): 055303.
13. Oliver-Ortega, Helena, Fernando Julian, Francesc X. Espinach, Quim Tarrés, Mònica Ardanuy, and Pere Mutjé. "Research on the use of lignocellulosic fibers reinforced bio-polyamide 11 with composites for automotive parts: Car door handle case study," *Journal of Cleaner Production* 226 (2019): 64–73.
14. Jariwala, Hitesh, and Piyush Jain. "A review on mechanical behavior of natural fiber reinforced polymer composites and its applications," *Journal of Reinforced Plastics and Composites* 38, no. 10 (2019): 441–453.
15. Jaafar, Jamiluddin, Januar Parlaungan Siregar, Salwani Mohd Salleh, Mohd Hazim Mohd Hamdan, Tezara Cionita, and Teuku Rihayat. "Important considerations in manufacturing of natural fiber composites: A review," *International Journal of Precision Engineering and Manufacturing-Green Technology* 6, no. 3 (2019): 647–664.
16. Baba, Zaid Ullah, Wani Khalid Shafi, Mir Irfan Ul Haq, and Ankush Raina. "Towards sustainable automobiles—advancements and challenges," *Progress in Industrial Ecology, an International Journal* 13, no. 4 (2019): 315–331.
17. Hassani, F. Oudrhiri, Nofel Merbahi, A. Oushabi, M. H. Elfadili, A. Kammouni, and N. Oueldna. "Effects of corona discharge treatment on surface and mechanical properties of Aloe Vera fibers," *Materials Today: Proceedings* 24 (2020): 46–51.
18. Rangappa, Sanjay Mavinkere, and Suchart Siengchin. "Lightweight natural fiber composites," *Journal of Applied Agricultural Science and Technology* 3, no. 2 (2019): 178.
19. Khan, Tabrej, Mohamed Thariq Bin Hameed Sultan, and Ahmad Hamdan Ariffin. "The challenges of natural fiber in manufacturing, material selection, and technology application: A review," *Journal of Reinforced Plastics and Composites* 37, no. 11 (2018): 770–779.
20. Li, Mi, Yunqiao Pu, Valerie M. Thomas, Chang Geun Yoo, Soydan Ozcan, Yulin Deng, Kim Nelson, and Arthur J. Ragauskas. "Recent advancements of plant-based natural fiber-reinforced composites and their applications," *Composites Part B: Engineering* 200 (2020): 108254.
21. Faruk, Omar, Jimi Tjong, and Mohini Sain, eds. *Lightweight and Sustainable Materials for Automotive Applications.* CRC Press, 2017.
22. Islam, Md Saiful, and Md Moynul Islam. "Sustainable reinforcers for polymer composites," In *Advances in Sustainable Polymer Composites*, pp. 59–88. Woodhead Publishing, 2021.
23. McGregor, Bruce A. "Physical, chemical, and tensile properties of cashmere, mohair, alpaca, and other rare animal fibers," In *Handbook of Properties of Textile and Technical Fibres*, pp. 105–136. Woodhead Publishing, 2018.
24. Gurukarthik Babu, B., D. Prince Winston, P. SenthamaraiKannan, S. S. Saravanakumar, and M. R. Sanjay. "Study on characterization and physicochemical properties of new natural fiber from Phaseolus vulgaris," *Journal of Natural Fibers* 16, no. 7 (2019): 1035–1042.

25. El-Sawalhi, R., J. Lux, and Patrick Salagnac. "Estimation of the thermal conductivity of hemp based insulation material from 3D tomographic images," *Heat and Mass Transfer* 52, no. 8 (2016): 1559–1569.
26. Soundhar, A., and Jayakrishna Kandasamy. "Mechanical, chemical and morphological analysis of crab shell/sisal natural fiber hybrid composites," *Journal of Natural Fibers* 18, no. 10 (2021): 1518–1532.
27. Dhakal, H. N., Z. Y. Zhang, R. Guthrie, James MacMullen, and Nick Bennett. "Development of flax/carbon fibre hybrid composites for enhanced properties," *Carbohydrate Polymers* 96, no. 1 (2013): 1–8.
28. Lotfi, Amirhossein, Huaizhong Li, Dzung Viet Dao, and Gangadhara Prusty. "Natural fiber-reinforced composites: A review on material, manufacturing, and machinability," *Journal of Thermoplastic Composite Materials* 34, no. 2 (2021): 238–284.
29. Ramesh, M., K. Palanikumar, and K. Hemachandra Reddy. "Mechanical property evaluation of sisal–jute–glass fiber reinforced polyester composites," *Composites Part B: Engineering* 48 (2013): 1–9.
30. Sanjay, M. R., P. Madhu, Mohammad Jawaid, P. Senthamaraikannan, S. Senthil, and S. Pradeep. "Characterization and properties of natural fiber polymer composites: A comprehensive review," *Journal of Cleaner Production* 172 (2018): 566–581.
31. Ilyas, R. A., and S. M. Sapuan. "Biopolymers and biocomposites: Chemistry and technology," *Current Analytical Chemistry* 16, no. 5 (2020): 500–503.
32. Zwawi, Mohammed. "A review on natural fiber bio-composites, surface modifications and applications," *Molecules* 26, no. 2 (2021): 404.
33. Tavares, Tânia D., Joana C. Antunes, Fernando Ferreira, and Helena P. Felgueiras. "Biofunctionalization of natural fiber-reinforced biocomposites for biomedical applications," *Biomolecules* 10, no. 1 (2020): 148.
34. Lotfi, A., H. Li, and D. V. Dao. "Natural fiber-reinforced composites: A review on material, manufacturing, and machinability," *Journal of Thermoplastic Composite Materials* 5 (2019): 1–47.
35. Dhakal, Hom N., Mikael Skrifvars, Kayode Adekunle, and Zhong Yi Zhang. "Falling weight impact response of jute/methacrylated soybean oil bio-composites under low velocity impact loading," *Composites Science and Technology* 92 (2014): 134–141.
36. Kandasamy, Jayakrishna, A. Soundhar, M. Rajesh, D. Mallikarjuna Reddy, and Vishesh Ranjan Kar. "Natural Fiber Composite for Structural Applications," In *Structural Health Monitoring System for Synthetic, Hybrid and Natural Fiber Composites*, pp. 23–35. Springer, Singapore, 2021.
37. Pickering, Kim L., M. G. Aruan Efendy, and Tan Minh Le. "A review of recent developments in natural fibre composites and their mechanical performance," *Composites Part A: Applied Science and Manufacturing* 83 (2016): 98–112.
38. Shahzad, Asim. "Hemp fiber and its composites–a review," *Journal of Composite Materials* 46, no. 8 (2012): 973–986.
39. Shen, Ziming, and Jiachun Feng. "Achieving vertically aligned SiC microwires networks in a uniform cold environment for polymer composites with high through-plane thermal conductivity enhancement," *Composites Science and Technology* 170 (2019): 135–140.
40. Friedrich, K. "Polymer composites for tribological applications," *Advanced Industrial and Engineering Polymer Research* 1, no. 1 (2018): 3–39.

41. Ulrich, Karl T. *Product Design and Development*. Tata McGraw-Hill Education, 2003.
42. Faruk, Omar, Andrzej K. Bledzki, Hans-Peter Fink, and Mohini Sain. "Progress report on natural fiber reinforced composites," *Macromolecular Materials and Engineering* 299, no. 1 (2014): 9–26.
43. Ita-Nagy, Diana, Ian Vázquez-Rowe, Ramzy Kahhat, Isabel Quispe, Gary Chinga-Carrasco, Nicolás M. Clauser, and María Cristina Area. "Life cycle assessment of bagasse fiber reinforced biocomposites," *Science of The Total Environment* 720 (2020): 137586.
44. Mansor, M. R., M. T. Mastura, S. M. Sapuan, and A. Z. Zainudin. "The environmental impact of natural fiber composites through life cycle assessment analysis," In *Durability and Life Prediction in Biocomposites, Fibre-Reinforced Composites and Hybrid Composites*, pp. 257–285. Woodhead Publishing, 2019.
45. Lee, Byoung-Ho, Hyun-Joong Kim, and Woong-Ryeol Yu. "Fabrication of long and discontinuous natural fiber reinforced polypropylene biocomposites and their mechanical properties," *Fibers and Polymers* 10, no. 1 (2009): 83–90.
46. Arshad, Muhammad, Manpreet Kaur, and Aman Ullah. "Green biocomposites from nanoengineered hybrid natural fiber and biopolymer," *ACS Sustainable Chemistry & Engineering* 4, no. 3 (2016): 1785–1793.
47. Siakeng, Ramengmawii, Mohammad Jawaid, Hidayah Ariffin, S. M. Sapuan, Mohammad Asim, and Naheed Saba. "Natural fiber reinforced polylactic acid composites: A review," *Polymer Composites* 40, no. 2 (2019): 446–463.
48. Mohanty, A. K., A. Wibowo, M. Misra, and L. T. Drzal. "Effect of process engineering on the performance of natural fiber reinforced cellulose acetate biocomposites," *Composites Part A: applied science and manufacturing* 35, no. 3 (2004): 363–370.
49. Ishikawa, Hiroki, Hitoshi Takagi, Antonio N. Nakagaito, Mikito Yasuzawa, Hiroaki Genta, and Hiroshi Saito. "Effect of surface treatments on the mechanical properties of natural fiber textile composites made by VaRTM method," *Composite Interfaces* 21, no. 4 (2014): 329–336.
50. Lopattananon, Natinee, Kuljanee Panawarangkul, Kannika Sahakaro, and Bryan Ellis. "Performance of pineapple leaf fiber–natural rubber composites: The effect of fiber surface treatments," *Journal of Applied Polymer Science* 102, no. 2 (2006): 1974–1984.
51. Demir, Hasan, Ulaş Atikler, Devrim Balköse, and Funda Tıhmınlıoğlu. "The effect of fiber surface treatments on the tensile and water sorption properties of polypropylene-luffa fiber composites," *Composites Part A: Applied Science and Manufacturing* 37, no. 3 (2006): 447–456.
52. Neto, J. S. S., R. A. A. Lima, D. K. K. Cavalcanti, J. P. B. Souza, R. A. A. Aguiar, and M. D. Banea. "Effect of chemical treatment on the thermal properties of hybrid natural fiber-reinforced composites," *Journal of Applied Polymer Science* 136, no. 10 (2019): 47154.
53. Gholampour, Aliakbar, and Togay Ozbakkaloglu. "A review of natural fiber composites: Properties, modification and processing techniques, characterization, applications," *Journal of Materials Science* 55, no. 3 (2020): 829–892.
54. Sood, Mohit, and Gaurav Dwivedi. "Effect of fiber treatment on flexural properties of natural fiber reinforced composites: A review," *Egyptian Journal of Petroleum* 27, no. 4 (2018): 775–783.

55. Sair, S., A. Oushabi, A. Kammouni, O. Tanane, Y. Abboud, F. Oudrhiri Hassani, A. Laachachi, and A. El Bouari. "Effect of surface modification on morphological, mechanical and thermal conductivity of hemp fiber: Characterization of the interface of hemp–Polyurethane composite," *Case Studies in Thermal Engineering* 10 (2017): 550–559.

56. De Araujo Alves Lima, Rosemere, Daniel Kawasaki Cavalcanti, Jorge De Souza e Silva Neto, Hector Meneses da Costa, and Mariana Doina Banea. "Effect of surface treatments on interfacial properties of natural intralaminar hybrid composites," *Polymer Composites* 41, no. 1 (2020): 314–325.

57. Neto, Jorge, Henrique Queiroz, Ricardo Aguiar, Rosemere Lima, Daniel Cavalcanti, and Mariana Doina Banea. "A Review of Recent Advances in Hybrid Natural Fiber Reinforced Polymer Composites," *Journal of Renewable Materials* 10, no. 3 (2022): 561.

58. Thiagamani, Senthil Muthu Kumar, Senthilkumar Krishnasamy, Chandrasekar Muthukumar, Jiratti Tengsuthiwat, Rajini Nagarajan, Suchart Siengchin, and Sikiru O. Ismail. "Investigation into mechanical, absorption and swelling behaviour of hemp/sisal fibre reinforced bioepoxy hybrid composites: Effects of stacking sequences," *International Journal of Biological Macromolecules* 140 (2019): 637–646.

59. Chaudhary, Vijay, Pramendra Kumar Bajpai, and Sachin Maheshwari. "Studies on mechanical and morphological characterization of developed jute/hemp/flax reinforced hybrid composites for structural applications," *Journal of Natural Fibers* 15, no. 1 (2018): 80–97.

60. Kumar, K. Senthil, I. Siva, N. Rajini, J. T. Winowlin Jappes, and S. C. Amico. "Layering pattern effects on vibrational behavior of coconut sheath/banana fiber hybrid composites," *Materials & Design* 90 (2016): 795–803.

61. Manjunath, V., and S. Udupa. "A study on hybrid composite using areca and eucalyptus fiber by using epoxy resin," *Journal of Mechanical and Industrial Engineering Research* 1 (2016): 1–2.

62. Mohanavel, Vinayagam, Thandavamoorthy Raja, Anshul Yadav, Manickam Ravichandran, and Jerzy Winczek. "Evaluation of mechanical and thermal properties of jute and ramie reinforced epoxy-based hybrid composites," *Journal of Natural Fibers* (2021): 1–11.

63. Abedom, Fasika, S. Sakthivel, Daniel Asfaw, Bahiru Melese, Eshetu Solomon, and S. Senthil Kumar. "Development of natural fiber hybrid composites using sugarcane bagasse and bamboo charcoal for automotive thermal insulation materials," *Advances in Materials Science and Engineering* 2021 (2021): 1–10.

64. Chandramohan, D., T. Sathish, S. Dinesh Kumar, and M. Sudhakar. "Mechanical and thermal properties of jute/aloevera hybrid natural fiber reinforced composites," In *AIP Conference Proceedings*, vol. 2283, no. 1, p. 020084. AIP Publishing LLC, 2020.

65. Atmakuri, Ayyappa, Arvydas Palevicius, Lalitnarayan Kolli, Andrius Vilkauskas, and Giedrius Janusas. "Development and analysis of mechanical properties of caryota and sisal natural fibers reinforced epoxy hybrid composites," *Polymers* 13, no. 6 (2021): 864.

66. Pereira, Alexandre L., Mariana D. Banea, Jorge S. S. Neto, and Daniel K. K. Cavalcanti. "Mechanical and thermal characterization of natural intralaminar hybrid composites based on sisal," *Polymers* 12, no. 4 (2020): 866.

67. Asim, Mohammad, Mohammad Jawaid, Md Tahir Paridah, Naheed Saba, Mohammed Nasir, and Rao M. Shahroze. "Dynamic and thermo-mechanical properties of hybridized kenaf/PALF reinforced phenolic composites," *Polymer Composites* 40, no. 10 (2019): 3814–3822.
68. Pappu, Asokan, Kim L. Pickering, and Vijay Kumar Thakur. "Manufacturing and characterization of sustainable hybrid composites using sisal and hemp fibres as reinforcement of poly (lactic acid) via injection moulding," *Industrial Crops and Products* 137 (2019): 260–269.
69. Arumugam, Soundhar, Jayakrishna Kandasamy, Mohamed Thariq Hameed Sultan, Ain Umaira Md Shah, and Syafiqah Nur Azrie Safri. "Investigations on fatigue analysis and biomimetic mineralization of glass fiber/sisal fiber/chitosan reinforced hybrid polymer sandwich composites," *Journal of Materials Research and Technology* 10 (2021): 512–525.
70. Baihaqi, N. M. Z. Nik, A. Khalina, N. Mohd Nurazzi, H. A. Aisyah, S. M. Sapuan, and R. A. Ilyas. "Effect of fiber content and their hybridization on bending and torsional strength of hybrid epoxy composites reinforced with carbon and sugar palm fibers," *Polimery* 66, no. 1 (2021): 36–43.
71. Ramasamy, Muthukumaran, Ajith Arul Daniel, M. Nithya, S. Sathees Kumar, and R. Pugazhenthi. "Characterization of natural–Synthetic fiber reinforced epoxy based composite–Hybridization of kenaf fiber and kevlar fiber," *Materials Today: Proceedings* 37 (2021): 1699–1705.
72. Hajiha H., Sain M. High toughness hybrid biocomposite process optimization, *Composites Science and Technology* 2015 May 6; 111:44–49.
73. Yahaya, R., S. M. Sapuan, M. Jawaid, Z. Leman, E. S. Zainudin. Effect of layering sequence and chemical treatment on the mechanical properties of woven kenaf-aramid hybrid laminated composites, *Materials & Design*. 2015 Feb 15;67:173–179.
74. Jothibasu, S., S. Mohanamurugan, R. Vijay, D. Lenin Singaravelu, A. Vinod, and M. R. Sanjay. "Investigation on the mechanical behavior of areca sheath fibers/jute fibers/glass fabrics reinforced hybrid composite for light weight applications," *Journal of Industrial Textiles* 49, no. 8 (2020): 1036–1060.
75. Sapuan, S. M., H. S. Aulia, R. A. Ilyas, A. Atiqah, T. T. Dele-Afolabi, M. N. Nurazzi, A. B. M. Supian, and M. S. N. Atikah. "Mechanical properties of longitudinal basalt/woven-glass-fiber-reinforced unsaturated polyester-resin hybrid composites," *Polymers* 12, no. 10 (2020): 2211.
76. Saba, Naheed, Paridah Md Tahir, and Mohammad Jawaid. "A review on potentiality of nano filler/natural fiber filled polymer hybrid composites," *Polymers* 6, no. 8 (2014): 2247–2273.
77. Rajesh, M., and Jeyaraj Pitchaimani. "Dynamic mechanical analysis and free vibration behavior of intra-ply woven natural fiber hybrid polymer composite," *Journal of Reinforced Plastics and Composites* 35, no. 3 (2016): 228–242.

Chapter 10

Data-driven optimization of manufacturing processes

T. Ramkumar
Dr. Mahalingam College of Engineering and Technology, Pollachi, India

M. Selvakumar and S. K. Ashok
Dr. Mahalingam College of Engineering and Technology, Pollachi, India

M. Mohanraj
Hindusthan College of Engineering and Technology, Coimbatore, India

CONTENTS

10.1 INTRODUCTION

In welding, detailed guidelines determine the quality of the final product. Following these guidelines minimizes time worked, reduces manpower, as well as material waste. Many researchers have studied various optimization techniques, some of which are discussed here.

Jamalian et al. studied the multi-pass tool pin varying FSW of AA5086-H34 plates reinforced with Al_2O_3 nanoparticles. The results were optimized using Artificial Neural Network (ANN) and improved the performance of tool

DOI: 10.1201/9781003252108-10

pin life [1]. Similarly, the tool parameter optimization was performed for AA7075-T651 and AA6061 aluminum alloys using the Taguchi method (Yuvaraj et al.) [2]. However, the relationship between the process parameters following that hardness of the ferritic steel joints was optimized through RSM by Güleryüz [3].

Palanivel, et al. predicted the optimization of wear resistance of friction stir welded dissimilar aluminum alloy (AA6351-T6 and AA5086-H111) with plates measured at 100 × 50 × 6 mm, and they tested the wear behavior using pin-on-disc apparatus (DUCOM TR20-LE) at room temperature according to ASTM G99-04 with a specimen size of 6 × 6 × 30 mm, sliding velocity of 1.5 m/s, force of 25N, and a sliding distance of 2500. They observed that wear resistance decreased with the increase in tool rotational speed [4].

Aruri et al. studied the wear and mechanical properties of AA 6061-T6 by friction stir processing and tested wear rate in the pin-on-disc tribometer as per ASTM G99-05 with prismatic pins of 8 mm diameter and sliding track of 100 mm with constant load 40 N, rotational speed 650 rpm, and a sliding speed of 3.4 m/s [5].

Varatharaju, et al. studied the wear behavior of an aluminum metal matrix composite and the sliding wear on the DUCOM pin-on-disc apparatus. A friction test was carried out between stationary pin stylus and a rotating disc by varying the load, speed, and track diameter as per ASTM G99 [6]. A wear test was conducted at a rotating speed of 200, 400, and 600 rpm over a range of applied loads for three minutes. The load was increased gradually from 1 to 3kg.

Koksal et al. studied the dry sliding wear behavior of in situ AlB2/Al composite based on the Taguchi method [7]. The wear tests were conducted on the pin-on-disc apparatus to determine the specific wear rate.

Extensive researches were conducted regarding aluminum weldments, but there are still some gaps, such as a variation of tool rotational speed and time duration of the weldments. The current study focused on two aspects—the first on various optimization techniques involved for FSW process, and the second being a case study of parameter optimization for AA2014-T6 similar weldments for tribological behavior. Based upon the detail literature survey, frequently and infrequently used optimization outcomes are also discussed.

10.2 IMPORTANCE OF MODELING AND OPTIMIZATION

Manufacturing industries strive for either a minimum cost of production or a maximum production rate—or an optimum combination of both—along with better product quality in machining.

Sahu et al. practiced grey relational analysis on friction stir welding for optimizing studies to fusion of different materials [8]. Sankar et al. applied

Taguchi modeling and optimization of friction stir welding on AA6061 alloy [9]. Authors experienced the device speediness as an additional substantial constraint than joining speed in FSW. In multi objective optimization, it is necessary that the factor is more important for deciding the performance of any process.

Jayaraj et al. conducted the electro chemical rust performance on welding region of FSW welded different joints of AA6061 aluminum-AZ31B magnesium alloys [10]. They discussed the diameter ratios, which are important in estimating the quality of welding joints. Darwins et al. noticed the best FSW responses by the Taguchi method, followed grade prediction and optimization of input values in FSW processed AM20 mg alloy [11]. Analysis of variance (ANOVA) is also used for finding the percentage contribution of FSW process parameter on multiple objectives. Taguchi method is used for evaluating the valuations and to find the grade with entropy rank calculation suitable for multi-objective optimization. Hardness and Tensile strength are the suitable parameters when predicting the appropriate welding solutions.

10.3 SOFT COMPUTING APPROACHES

10.3.1 Artificial neural network (ANN)

Naik et al. studied the mechanical properties of stainless steel 2205 joining using the GTAW method [12]. Using ANN back propagation algorithm, the parameters of the welding was optimized. Variation of the process parameters was detected using ANOVA. Following that, the optimal level of factors was verified by confirmation test. In this case, current, time speed, variation of oxide fluxes, and electrode diameter and gas flow rate were chosen as the parameters.

Muthu Krishnan et al. investigated the mechanical properties of AA6063 and A319 joining using FSW and evaluated the process parameter optimization through RSM and ANN [13]. Using these techniques, the error rate was predicted to be low in regression ANN, compared to RSM.

10.3.2 Full factorial method

Prakash et al. examined the influence of process parameters of aluminum alloy through full factorial design [14]. The author selected the following process parameters such as tool speed (rpm), feed rate (mm/min), displacement (mm/min), and welding time (min). Impact strength was chosen as response for this analysis, and the quality of the welded joints was also analyzed.

Aita et al. analyzed the shear strength of AA6060-T5 weldments through full factorial design [15]. The welding parameters were optimized by two different methodologies such as Taguchi and full factorial design. Three different parameters were chosen such as speed, plunge time, and dwell time. The joint strength determined by the welding parameters was obtained by quadratic regression equation. By using this methodology, the welding parameters were within acceptable limits and also provided optimum welding parameters.

10.3.3 Response surface methodology

Subramanian et al. studied the optimization of the parameters in the friction stir welding process of magnesium using RSM [16]. Three different tool profiles (square, conical, and cylindrical), three different rotational speeds (900, 1150 and 11400 rpms), and three different traverse speeds (25, 35, and 45 mm/mins) were used. The Box-Behnken model was used to evaluate the optimum parameters. Using this technique, the optimum tensile strength was 186.721 Mpa, and the tool rotation speed was 900 rpm, with the square tool pin head of 45 mm tool traverse speed predicted as optimum level. Moreover, the test results showed the tensile strength was 182.57 Mpa and the percentage of error was 2.2 %. It was very closely associated with the experimental values.

10.3.4 Technique for order of preference by similarity to ideal solution (TOPSIS)

Banik et al. determined the better tool geometry for FSW weldments for AA 6061-T6 using the hybrid PCA-TOPSIS optimization method [17]. The mechanical properties—ultimate tensile strength, yield strength, elongation, and hardness—were optimized using TOPSIS. The process parameters such as rotational speed of tool and traverse speed of tool were taken. From this analysis, it can be concluded that tool geometries of taper threaded mode provide better properties. The confirmation test also determined the optimized parameters were within acceptable limits.

Mostafa AKBARI et al. optimized the mechanical properties of $B_4C/A356$ composites processed by FSP using hybrid multi-objective optimization such as TOPSIS and modified NSGA-II. FSP parameters such as rotational speed, traverse speed, and tool pin profile affected the base metals [18].

10.4 OPTIMIZATION OF MANUFACTURING PROCESS

Manufacturing composites is broadly classified as solid metallurgy and liquid metallurgy. Akbarpour et al. studied the influence of CNT in aluminum composites through the hot pressing technique [19]. The results clearly showed that increasing the CNT vol% decreased the grain size of the matrix. Reinforcing 4 vol% of CNT increased the yield strength of the composites. Additionally, the author illustrates that grain size is one of the most influencing factors for increasing mechanical properties.

Turan et al. investigated the mechanical and tribological properties of magnesium composites reinforced with fullerene by semi powder metallurgy techniques, and the results indicated that 0.5 wt% addition of fullerene shows better mechanical and wear behavior over unreinforced magnesium [20]. The rate of corrosion is also increased with the addition of fullerene.

Dixit et al. deliberated the effect of copper granules and also studied the interfacial bonding with a copper-graphite composite fabricated by flake powder metallurgy [21]. Mechanical milling was used to develop the flake-like composite particles. The copper laminates increased interfacial adhesion during sintering. Mechanical milling dispersed the secondary particles in the matrix, provided the better interfacial adhesion, and improved the physical properties of the composites.

Malin Chen et al. studied the heat treatment of Al 6061 reinforced with CNT using the powder metallurgy technique, and results showed that the addition of CNT weakened the hardening of Al alloy [22]. CNT provided synergy with the precipitation phase, which increased the hardening capacity of the composites.

Wei et al. assessed the microstructure and mechanical properties of CNT-reinforced molybdenum-hafnium-carbon composites through powder metallurgy route, which showed improved results of even dispersion of secondary particles [23]. Mo-based composites at 30 h of milling provide better density, microstructure, and mechanical properties.

Mohanavel et al. deliberated the microstructure and mechanical properties of AA 7075 reinforced with hard ceramic particles through the liquid metallurgy technique, and the results revealed that hardness and the mechanical properties such as mechanical and tribological properties of the composites were increased because of the even distribution of the secondary particles [24].

Sadhasivam et al. enhanced the mechanical and thermal properties of AISI 420 reinforced with TiB_2 by the liquid metallurgy technique, and the results indicated that extreme tensile strength was obtained because of uniform distribution of hard ceramic particles [25]. However, ductile to brittle transition was observed, while the TiB_2 content increased.

Gangadhar et al. evaluated the mechanical properties of Al7029-flyash-WC hybrid composites using liquid metallurgy. The results illustrated that no significant change was observed during impact strength, and also that many dimples were observed in the surfaces during experimentation [26].

10.5 CASE STUDY

The friction stir welded AA2014-T6 specimen was subjected to a wear test. Input parameters such as tool rotational speed (TRS), welding speed (WS), and axial force (AF) were selected, and the output response such as

Figure 10.1 Pin-on-disc wear testing machine.

average mass loss (g), wear rate (mm³/m), and specific wear rate (mm³/Nm) were examined. Wear testing parameters such as applied load (20 N), sliding speed (1.5 m/s), track diameter (100 mm), and sliding distance (1500 m) were kept constant. The optimum welding parameter to achieve better response was investigated using a Taguchi analysis.

10.5.1 Wear test apparatus

Wear tests were performed by a DUCOM pin-on-disc machine, as shown in Figure 10.1. This machine assesses the wear and friction characteristics of materials subjected to sliding contacts in dry or lubricated environments. The wear test is conducted by rotating the disc against a stationary pin and varying the load applied, disc speed, and track diameter.

10.5.2 Wear test parameters

The welded specimen, made into a pin of 3 mm square and 30 mm height, is used for testing. The disc of 12 mm thickness and 150 mm dia. is used as counterpart. The pin is pressed against the disc with a load of 20 N. The sliding speed of 1.5 m/s, a track dia. of 100 mm, and sliding distance of 1500 m are selected for the test. Wear rate in mm³/min is determined using the weight loss method.

The wear rate of the samples was calculated from the average mass losses using the formula as given in Equation 10.1.

$$\text{Wear Rate, WR} = \frac{\text{ML} * 1000}{\rho * \text{SD}}$$

(10.1)

where,
ML = Mass loss in cm³
ρ = Density (2.8 g/cm³) and
SD = Sliding distance (1500 m)

The specific wear rates of the samples were calculated from the average mass losses using the formula as given in Equation 10.2.

$$\text{Specific Wear Rate} = \frac{WR}{L} \tag{10.2}$$

where,
WR = Wear rate
L = Load applied (20 N)

10.5.3 Experimental results

The test surface of the wear specimens was made flat by mechanical polishing and cleaned using acetone. The wear surface of the samples and wear track are cleaned using acetone and weighed using a microbalance before and after each test. The microbalance has an accuracy of +0.0001 gm. The wear samples are held against the rotating disc by applying the load and adjusting the sliding speed. The procedure is repeated for all the samples by keeping the test parameters constant. The calculated mass loss, wear rates, and specific wear rates of the samples are shown in Table 10.1.

It is inferred that wear rate varies significantly with the variation in weld parameters. The wear rate of the specimens welded with low tool rotation speed (700 rpm) is between 0.0020 mm³/m to 0.0035 mm³/m. The wear rate gradually decreases with the increasing weld speed for the low speed of the

Table 10.1 Mass loss, wear rate, and specific wear rate of the specimens

Sample	Average mass loss (g)	Wear rate (mm³/m)	Specific wear rate (mm³/Nm)
S1	0.0142	0.003381	0.000169
S2	0.0121	0.002881	0.000144
S3	0.0092	0.002190	0.000110
S4	0.0138	0.003286	0.000164
S5	0.0091	0.002167	0.000108
S6	0.0188	0.004476	0.000224
S7	0.0193	0.004595	0.000230
S8	0.0164	0.003905	0.000195
S9	0.0175	0.004167	0.000208

tool. For the samples welded with high tool rotation speed (1400 rpm), the wear rates are higher compared to the other samples. The highest wear rate is observed for the sample S7 (0.004595 mm³/m). The wear rate is observed as the lowest for the sample S5 (0.002167 mm³/m). There is no trend observed for the wear rate for the samples welded with tool rotation speeds of 1000 and 1400 rpm (i.e., samples S4 to S9). Hence, a Taguchi analysis is performed through MINITAB software to analyze the consequence of welding parameters on the wear rate.

10.5.4 Taguchi analysis

The Taguchi analysis is performed to study the effect of weld process parameters on the wear rate of the samples. The smaller the better rule is selected, since the wear rate of the joint should be smaller. The plots for signal-to-noise (S/N) ratio and means are illustrated in Figures 10.2 and 10.3, respectively.

It is found from the analysis that the tool rotational speed is the major factor in controlling the wear rate, followed by plunge depth, and then the weld speed. The response table for the signal-to-noise ratio is shown in Table 10.2. Analysis of Variance (ANOVA) for wear rate using Sequential Sum of Squares is shown in Table 10.3.

It is observed from the table that the factor tool rotation speed is the statistically major influencing factor in controlling the wear rate compared to the weld speed and the axial force. The maximum contribution is 45.69% by the tool rotation speed [27, 28]. The next highest contribution is 19.72%

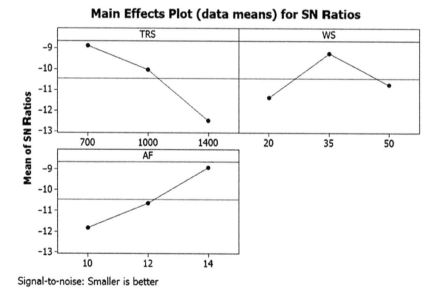

Main Effects Plot (data means) for SN Ratios

Signal-to-noise: Smaller is better

Figure 10.2 Main effects plot for SN ratio of wear rate.

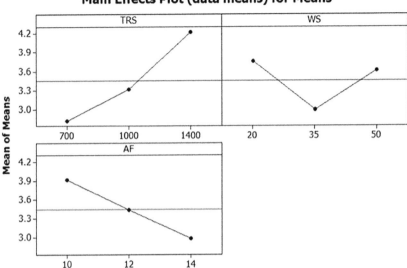

Main Effects Plot (data means) for Means

Figure 10.3 Main effects plot for means of wear rate.

Table 10.2 Response table for S/N ratios for wear rate

Level	TRS	WS	AF
I	−8.860	−11.387	−11.810
2	−10.023	−9.247	−10.640
3	−12.492	−10.741	−8.924
Rank	I	3	2

Table 10.3 ANOVA results for wear rate

Source	DoF	Seq SS	Adj SS	Adj MS	% contribution
TRS	2	3.0494	3.0494	1.5247	45.69
WS	2	1.0055	1.0055	0.5028	15.07
AF	2	1.3161	1.3161	0.6581	19.72
Error	2	1.3029	1.3029	0.6514	19.52
Total	8	6.6739	6.6739	3.337	100

by the axial force. The least influencing factor is the weld speed, with a contribution of 15.07%. A confirmation experiment is conducted by keeping the parameters at predicted optimum values of 700 rpm, 35 mm/min, and 14 kN for TRS, WS, and AF, respectively [29–31]. Three wear samples are cut from the welded plate, and the samples were tested for wear rate. The average wear rate is 1.935×10^{-3} mm³/m.

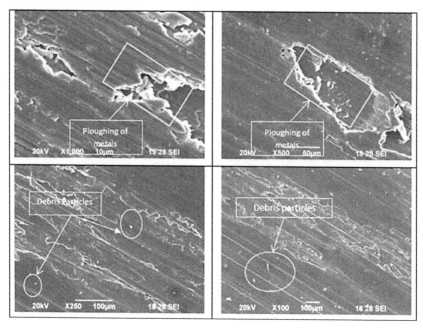

Figure 10.4 Wear morphology of sample S1.

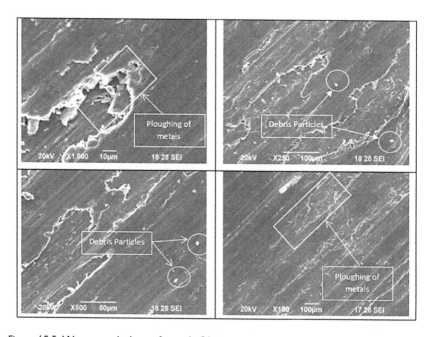

Figure 10.5 Wear morphology of sample S6.

The SEM images of the wear samples S1 and S6 after the dry sliding wear test are shown in Figures 10.4 and 10.5, respectively. The SEM images of the sample S1 reveal the presence of small size wear debris particles, indicating a lesser wear rate. The SEM images of the sample S6 reveal the presence of large size wear debris particles, indicating a high wear rate [32].

10.6 CONCLUSION

In this chapter, a brief overview of the friction stir welding process parameter optimization using various techniques is deliberated in detail. From experimental studies presented regarding wear behavior of AA2014-T6, similar weldments and its process parameters are optimized using Taguchi techniques. The result reveals that the wear rate gradually increases with the increasing rotation speed of 700 rpm. For the samples welded with high tool rotation speed (1400 rpm), the wear rates are higher compared to the other samples. It is observed from the ANOVA results that the rotation speed is controlling the wear rate compared to the weld speed and the axial force.

REFERENCES

1. Hasan Mohammadzadeh Jamalian, Mehran Tamjidi Eskandar, Amir Chamanara, Reza Karimzadeh, Razieh Yousefian (2021), "An artificial neural network model for multi-pass tool pin varying FSW of AA5086-H34 plates reinforced with Al_2O_3 nanoparticles and optimization for tool design insight," *CIRP Journal of Manufacturing Science and Technology*, 35, 69–79.
2. K. P. Yuvaraj, P. A. Varthanan, L. Haribabu, R. Madhubalan, K. P. Boopathiraja (2021), "Optimization of FSW tool parameters for joining dissimilar AA7075-T651 and AA6061 aluminium alloys using Taguchi Technique," *Materials Today: Proceedings*, 45(2), 919–925.
3. Güldal Güleryüz (2020), "Relationship between FSW parameters and hardness of the ferritic steel joints: Modeling and optimization," *Vacuum*, 178, 109449.
4. R. Palanivel, P. K. Mathews, N. Murugan, I. Dinaharan (2012), "Prediction and optimization of wear resistance of friction stir welded dissimilar aluminum alloy," *Procedia Engineering*, 38, 578–584.
5. Devaraju Aruri, Kumar Adepu, Kumaraswamy Adepu, Kotiveerachari Bazavada (2013), "Wear and mechanical properties of 6061-T6 aluminum alloy surface hybrid composites [(SiC + Gr) and (SiC + Al_2O_3)] fabricated by friction stir processing," *Journal of Materials Research and Technology*, 2, 362–369.
6. S. Varatharaju, T. P. Ramaswamy, B. Sreedhara (2014), "The abrasive wear behavior of cenosphere-aluminium metal matrix composite," *Acta Metallurgica Slovaca*, 20(2), 177–188.
7. S. Koksal, F. Ficici, R. Kayikci, O. Savas (2012), "Experimental optimization of dry sliding wear behavior of in situ AlB2/Al composite based on Taguchi's method," *Materials & Design*, 42, 124–130.

8. Prakash Kumar Sahu, Sukhomay Pal, (2015), "Multi-response optimization of process parameters in friction stir welded AM20 magnesium alloy by Taguchi grey relational analysis," *Journal of Magnesium and Alloys*, 3, 36.

9. B. Ravi Sankar, Uma Maheswar Rao (2017), "Modelling and Optimisation of Friction Stir Welding on AA6061 Alloy," *Materials Today: Proceedings*, 4, 7448–7456.

10. R. Kamal Jayaraj, S. Malarvizhi, V. Balasubramanian (2017), "Electrochemical corrosion behaviour of stir zone of friction stir welded dissimilar joints of AA6061 aluminium–AZ31B magnesium alloys," *Transaction of Nonferrous Metals Society of China*, 27, 2181.

11. A. K. Darwins, M. Satheesh, G. Ramanan (2018), "Modelling and optimization of friction stir welding parameters of Mg-ZE42 alloy using grey relational analysis with entropy measurement," *IOP Conference Series: Materials Science and Engineering* 402, 012162.

12. A. Balaram Naik, A. Chennakeshava Reddy (2018), "Optimization of tensile strength in TIG welding using the Taguchi method and analysis of variance (ANOVA)," *Thermal Science and Engineering Progress*, 8, 327–339.

13. M. Muthu Krishnan, J. Maniraj, R. Deepak, K. Anganan (2018), "Prediction of optimum welding parameters for FSW of aluminium alloys AA6063 and A319 using RSM and ANN," *Materials Today: Proceedings*, 5(1), 716–723.

14. M. Prakash, A. Daniel Das (2021), "Investigation on effect of FSW parameters of aluminium alloy using Full Factorial Design," *Materials Today: Proceedings*, 37(2), 608–613.

15. C. A. G. Aita, I. C. Goss, T. S. Rosendo, M. D. Tier, A. Wiedenhöft, A. Reguly (2020), "Shear strength optimization for FSSW AA6060-T5 joints by Taguchi and full factorial design," *Journal of Materials Research and Technology*, 9(6), 16072–16079.

16. M. Subramanian, N. Sathishkumar, Joses Samuel Sasthrigal, N. Arunkumar, V. Hariharan (2020), "Optimization of process parameters in friction stir welded dissimilar magnesium alloys using RSM," *Materials Today: Proceedings*, doi:10.1016/j.matpr.2020.09.049.

17. Abhijit Banik, Abhijit Saha, John Deb Barma, Uttam Acharya, Subhash Chandra Saha (2021), "Determination of best tool geometry for friction stir welding of AA 6061-T6 using hybrid PCA-TOPSIS optimization method," *Measurement*, 173, 108573.

18. Mostafa Akbari, Mohammad Hasan Shojaeefard, Parviz Asadi, Abolfazl Khalkhali (2017), "Hybrid multi-objective optimization of microstructural and mechanical properties of B4C/A356 composites fabricated by FSP using TOPSIS and modified NSGA-II," *Transactions of Nonferrous Metals Society of China*, 27, 2317–2333.

19. M. R. Akbarpour, A. Pouresmaeil (2018), "The influence of CNTs on the microstructure and strength of Al-CNT composites produced by flake powder metallurgy and hot pressing method," *Diamond and Related Materials*, 88, 6–11.

20. Muhammet Emre Turan, Yavuz Sun, Yasin Akgul (2018), "Mechanical, tribological and corrosion properties of fullerene reinforced magnesium matrix composites fabricated by semi powder metallurgy," *Journal of Alloys and Compounds*, 740, 1149–1158.

21. Manish Dixit, Rajeev Srivastava (2019), "The effect of copper granules on interfacial bonding and properties of the copper-graphite composite prepared by flake powder metallurgy," *Advanced Powder Technology*, 30, 3067–3078.

22. Malin Chen, Genlian Fan, Zhanqiu Tan, Chao Yuan, Qiang Guo, Dingbang Xiong, Mingliang Chen, Quan Zheng, Zhiqiang Li, Di Zhang (2019), "Heat treatment behavior and strengthening mechanisms of CNT/6061Al composites fabricated by flake powder metallurgy," *Materials Characterization*, 153, 261–270.
23. Yong Wei, Lai-Ma Luo, Huai-Bing Liu, Xiang Zan, Jiu-Peng Song, Qiu Xu, Xiao-Yong Zhu, Wu Yu-Cheng (2020), "A powder metallurgy route to fabricate CNT-reinforced molybdenum-hafnium-carbon composites," *Materials & Design*, 191, 108635.
24. V. Mohanavel, S. Suresh Kumar, T. Sathish, T. Adithiyaa, K. Mariyappan (2018), "Microstructure and mechanical properties of hard ceramic particulate reinforced AA7075 alloy composites via liquid metallurgy route," *Materials Today: Proceedings*, 5, 26860–26865.
25. Vikram Kumar Sadhasivam S. Jain (2021), "Enhanced mechanical and thermal properties of AISI 420/TiB2 composites fabricated by liquid metallurgy route," *Composites Communications*, 23, 100550.
26. T. G. Gangadhar, D. P. Girish, J. Satheesh, H. Govtham, K. S. Pradeep (2021), "Evaluation of mechanical properties of Al7029/flyash/WC hybrid composites developed by liquid metallurgy," *Materials Today: Proceedings*, 46, 5969–5974.
27. C. A. G. Aita, I. C. Goss, T. S. Rosendo, M. D. Tier, A. Wiedenhöft, A. Reguly (2020), "Shear strength optimization for FSSW AA6060-T5 joints by Taguchi and full factorial design," *Journal of Materials Research and Technology*, 9(6), 16072–16079.
28. M. Subramanian, N. Sathishkumar, Joses Samuel Sasthrigal, N. Arunkumar, V. Hariharan (2020), "Optimization of process parameters in friction stir welded dissimilar magnesium alloys using RSM," *Materials Today: Proceedings*, doi:10.1016/j.matpr.2020.09.049.
29. Malin Chen, Genlian Fan, Zhanqiu Tan, Chao Yuan, Qiang Guo, Dingbang Xiong, Mingliang Chen, Quan Zheng, Zhiqiang Li, Di Zhang (2019), "Heat treatment behavior and strengthening mechanisms of CNT/6061Al composites fabricated by flake powder metallurgy," *Materials Characterization*, 153, 261–270.
30. Manish Dixit, Rajeev Srivastava (2019), "The effect of copper granules on interfacial bonding and properties of the copper-graphite composite prepared by flake powder metallurgy," *Advanced Powder Technology*, 30, 3067–3078.
31. M. R. Akbarpour, A. Pouresmaeil (2018), "The influence of CNTs on the microstructure and strength of Al-CNT composites produced by flake powder metallurgy and hot pressing method," *Diamond and Related Materials*, 88, 6–11.
32. V. Mohanavel, S. Suresh Kumar, T. Sathish, T. Adithiyaa, K. Mariyappan (2018), "Microstructure and mechanical properties of hard ceramic particulate reinforced AA7075 alloy composites via liquid metallurgy route," *Materials Today: Proceedings*, 5, 26860–26865.

Chapter 11

Conclusion and challenges

G. Velmurugan and R. Sundarakannan

Saveetha School of Engineering, SIMATS, Sriperumbudur, India

CONTENTS

DOI: 10.1201/9781003252108-11

11.1 PROS AND CONS OF LIGHTWEIGHT MATERIALS

Light weighting has a favorable environmental impact since it uses fewer raw resources in the manufacturing process. As a result of light weighting, manufacturers have saved millions of dollars and considerably decreased greenhouse gas emissions caused by product production and shipping. Mining has a lower environmental impact, uses less energy, and produces less landfill waste. However, there are several unintended consequences of light weighting that are sometimes neglected. Modern autos and aerospace require advanced materials to improve fuel efficiency while retaining performance and safety. Lightweight materials have a lot of potential for improving vehicle efficiency because it requires less energy to accelerate a lighter item than a heavy one. A 10% weight decrease in a vehicle can improve fuel economy by 6% to 8%. For example, a 20% weight reduction on the Boeing 787 resulted in a 10% to 12% increase in fuel economy. In addition to lowering carbon emissions, lightweight design can improve flying performance by allowing for faster acceleration, increased structural strength and rigidity, and improved safety.

The use of lightweight materials such as high strength steel, magnesium (Mg) alloys, aluminum (Al) alloys, carbon fiber, and polymer composites to replace traditional steel components can reduce the weight of a vehicle's body and chassis by up to 50%, lowering its fuel costs. By 2030, using lightweight components and high-efficiency engines made possible by new materials in a fourth of the US fleet may save more than 5 billion gallons of petroleum each year.

11.1.1 High strength steel

Steel is now the most widely utilized construction material in various industries due to its relative ease of manufacture and availability, relatively high stiffness in the form of high strength steels, good dimensional qualities at high temperatures, and low cost among commercial aerospace materials. However, because of their high density and other drawbacks, such as a high susceptibility to corrosion and embrittlement, high strength steels are not widely used in aircraft products and parts. Steel typically makes up around 5% to 15% of the structural weight of passenger planes, with the percentage continuously reducing. Notwithstanding these drawbacks, high strength steels are still the preferred material for security components to ensure exceptional strength and stiffness.

Gearing, bearings, and undercarriage applications are the most common uses for high strength steels in aerospace. This type of steel could lower the weight of automobiles by up to 25%, especially in strength-limited intentions like door rings and pillars. It is generally compatible with existing manufacturing and materials currently used in vehicles.

Pros: corrosion performance, high strength, formability, and stiffness, as well as low cost

Cons: As strength increases, ductility declines, causing problems with forming and connecting, as well as wearing out stamping molds faster than with lower grades. Design, component processing, and behavior in hostile settings are all challenges.

11.1.2 Aluminum and its alloy

Aluminum's properties and processing are well understood by scientists due to its application in aerospace and engineering. It's currently used in vehicle panels and powertrain components, but it's expensive and difficult to fabricate. When combining aluminum with other materials, manufacturers confront challenges with joining, corrosion, maintenance, and recycling. Aluminum is increasingly being used for hoods, trunk lids, and doors as a lighter, more expensive alternative to steel. It can also reduce weigh considerably, up to 60%.

Despite the growing interest in high-performance composites like carbon fiber, aluminum alloys account for a large amount of aeronautical structural weight. Advanced aluminum alloys are a desirable lightweight material in many aerospace structural applications due to their high specific strength and stiffness, outstanding ductility and corrosion resistance, relatively cheap cost, and good manufacturability and reliability. By altering compositions and heat treatment procedures, aluminum alloys can provide a wide range of material qualities to fulfill a variety of application needs.

Pros: The technology is truly cutting-edge, with excellent strength, stiffness, and energy dissipation.

Cons: Costlier than steel, more difficult to combine with other materials, and limited formability.

11.1.3 Carbon fiber composites

Carbon fiber (CF), sometimes called Graphite Fiber, is made up of long strands of carbon fibers that are intertwined to produce a fabric-like pattern. Carbon fiber parts have characteristics similar to steel and a weight similar to plastic. Carbon fiber parts have a significantly greater strength-to-weight ratio (as well as stiffness-to-weight ratio) than steel or plastic parts. Carbon fiber is a strong composite material utilized as a reinforcing material. Global demand for CF will continue to increase, surpassing 77,000 tonnes in 2018 and topping 150,000 tonnes by 2025. In 2018, several manufacturers raised capacity, with Kangde (66,000 tonnes), Toray (62,000 tonnes), and SGL Group (49,000 tonnes) all planning significant expansion in the next few years. Hexcel, DowAska, and Hyosung also have expansion ambitions.

Pros: Carbon fiber composites are unique for many reasons. Here are a few examples:

- Lightweight – Carbon fiber is a low-density, high strength-to-weight ratio material.
- High tensile strength – Carbon fiber is one of the most difficult fibers to stretch or bend of all the commercial reinforcing fibers when it tensions load.
- Low thermal expansion – Carbon fiber expands and contracts significantly less than steel and aluminum in hot and cold temperatures.
- Exceptional durability – Carbon fiber has better fatigue characteristics than metal, which means that components constructed of carbon fiber will last longer under continual usage.
- Corrosion-resistance – Carbon fiber is one of the most corrosion-resistant materials available when produced with the right polymers.

Cons:

- Carbon fiber will break or shatter when it is crushed, pushed past its strength limits, or subjected to a significant level of impact. If you hit it with a hammer, it will fracture. Machining and perforations can also produce weak spots, increasing the risk of breaking.
- Carbon fiber is an elevated material that comes at a hefty cost. Although prices have dropped greatly in the last five years, demand has not increased sufficiently to allow for a considerable expansion in supply. As a result, costs are expected to remain stable in the coming years.

11.1.4 Magnesium and its alloy

Magnesium is currently utilized in powertrain castings and subassembly closures; however, several technological hurdles limit the use of magnesium in automotive applications. Even though magnesium (Mg) may lower component weight by more than 60%, its usage in vehicles is now limited to less than 1% of the total weight. Despite the impossibility of incorporating multiple, individually cast, or wrought Mg components into articulated sub-assemblies in the near future, Mg will continue to play a role in vehicle light weighting due to its appealing properties of low density, high specific stiffness, and amenability to thin-wall die casting and component implementation.

Pros: High rigidity and strength, and compatibility with current stamping infrastructure

Cons: Expensive and lacking in significant numbers from US producers to fulfill automobile demands. Ductility, joining, repair, recycling, and corrosion are among of the other problems. Rare earth additions may also be required to fulfill crash standards by improving energy absorption [1].

11.1.5 Polymer composite materials

Composite materials are frequently utilized in lightweight structural aviation and automobiles and with good reason: their unique characteristics enable engineers to overcome design challenges that would otherwise be difficult to resolve. Fiberglass, carbon fiber, and fiber-reinforced matrix systems are examples of common composite materials. The most prevalent is fiberglass, which was initially widely employed in boats and vehicles in the 1950s, the same decade that Boeing began using it in passenger aircraft. Composite materials make about 50% to 70% of today's airplane construction. Though composite materials offer several benefits in aviation, some critics believe they represent a safety concern. In this article, we'll go through the most important benefits and drawbacks of composite materials in aviation.

Pros: The fact that composite materials are lightweight is their biggest benefit. Composite materials may significantly reduce an aircraft's weight, resulting in enhanced performance and fuel economy. In most airplanes, fiber-reinforced matrix technologies outperform standard aluminum by providing a smoother, more aerodynamic surface that increases performance and fuel economy. Composite materials do not corrode as quickly as other construction types, and they do not break as easily as aluminum due to metal fatigue. Instead, they flex, allowing them to endure longer than metal and cost less to maintain and repair.

Cons: The most significant drawback of composite materials is that they are difficult to break. This may sound oxymoronic, but it just means that it is impossible to detect if the aircraft's internal structure has been harmed. It is easier to discover a need for repairs with aluminum since it bends and dents more quickly. Composite materials are also more difficult and expensive to repair than metals; however, the long-term advantages of choosing a more durable material can be argued to offset this expense [2].

11.2 CHALLENGES OF MAKING LIGHTWEIGHT MATERIALS

Light weighting has a number of advantages that will almost certainly lead to increasing usage of lightweight metals and other materials. However, manufacturers must change their design, production, maintenance, and repair procedures as part of the light-weighting transition. However, light weighing is difficult to accomplish in the production process. Employees in manufacturing, maintenance, and repair companies must adapt their operations to meet particular lightweight metal and other material needs as the demand for lightweight metal and other materials grows. The effects of light weighting on downstream production can be considerable.

Different processes: This has an effect on joining in particular. Welding aluminum to steel or other aluminum parts, for example, is faster than riveting and reduces the weight of the rivets, but it's still in the experimental stage and uses more energy than riveting, so it's only used in low-volume manufacturing. We've never utilized lightweight alloys in industrial development before, so we won't know exactly what adjustments to production lines they'll require until they're employed in real-world mass production.

Higher costs: Light weighting often raises manufacturing costs independent of material costs when compared to heavier metals, due to expenses involved with modifying production lines to suit light weighting mixed with learning curve labor costs when beginning to deal with lightweight materials. As a result, manufacturers are focusing light weighting on luxury items, which limits the market for lightweight products.

Differing performance: According to the Autodesk article, lightweight materials outperform heavy materials in many areas, but they also have drawbacks, such as poor ride and handling in vehicles due to lightweight materials' inclination to vibrate more than heavier materials. Parts that are excessively "flimsy" must be reinforced, increasing the cost and complexity of manufacture.

Safety concerns: If manufacturers save weight by sacrificing structural safety, the result may be a dangerous car, aircraft, or light-rail train. Light weight's increase may be limited due to safety considerations.

Maintenance and repair: When employing lightweight metals and other materials, designers must not only address manufacturing difficulties, but they must also prepare for essential modifications in lightweight product maintenance and repair—areas that will necessitate new procedures [3].

11.2.1 Challenges faced in Advanced High Strength Steel

Because of their contribution to efficient, mass-optimized vehicle constructions, Advanced High Strength Steel (AHSS)—including dual phase, TRIP, and martensitic grades—are being used in modern vehicle programmers. Increased use of advanced high strength steels improves the energy absorption, durability, and structural strength of body structures. However, several production obstacles must be overcome to properly employ these materials. Because these materials are more formable than traditional high strength steels, forming them presents a new difficulty. Increased press force and energy are required for successful AHSS stamping, which has an impact on the forming methods and die materials used. Engineers must additionally deal with the components' enhanced springback.

To accurately anticipate springback, advances in simulation technologies are necessary. In addition to the normal strain-based failures found in stamping operations, simulation technology is necessary to forecast stress-based failures. The characteristics of these materials necessitate higher amounts of alloying than are normally utilized in automobile body construction materials, making joining them difficult. The heat input of welding methods has an effect on multiphase microstructures. Increased hardenability also adds to the difficulty of spot-welding procedures [4].

11.2.2 Challenges faced in aluminum and its alloy

Aluminum is found in combination with other minerals, the most common of which is bauxite, rather than as an independent ore (it is too reactive with other compounds). Bauxite is mostly extracted from open pits in a broad area near the equator. Aluminum hydroxide, iron oxide, titanium dioxide, and kaolinite make up bauxite. Before it can be electrolyzed into aluminum, bauxite is treated with aluminum oxide (alumina). Alumina refineries use the Bayer process, which involves releasing alumina from bauxite in a digester using a caustic soda solution. A tonne of alumina is made from 2.9 tons of bauxite. The bauxite residue ("red mud") that results is made up of insoluble particles.

The storage and handling of bauxite waste from the Bayer process has long been a concern. A total of 120-million tonnes of residue is produced each year by the industry's 100 processors. Most of it is held in holding ponds, and only a small portion of the waste gets recycled. The largest challenge in the sector is the amount of energy utilized. Aluminum is jokingly referred to as "congealed electricity" due to its high consumption. Aluminum requires 211 GJ per tonne of energy to produce, whereas steel requires 22.7 GJ per tonne. The strength of the chemical connection between aluminum and oxygen is much stronger than the identical interaction between iron and oxygen, which explains the large amount of energy used. As a result, it takes a lot more energy to break the bond and produce the metal [5].

Because of economic benefits such as light weight, strong corrosion resistance, high toughness, severe temperature capabilities, and simple recyclability, aluminum and its alloys are widely utilized in construction and the aerospace and automotive industries. Welding of aluminum alloys is required for structural and mechanical fabrications such as airplanes. Welding, on the other hand, has its drawbacks and can be difficult. Porosity, hot cracking, partial fusion, and other welding faults are prevalent in aluminum.

11.2.3 Challenges faced in polymer composites

A wide range of manufacturing faults can occur during the manufacture of composite laminates at various stages. Defects such as resin-rich, resin-starved, or excessive porosity can arise during composite material production. Blisters, delamination, marcelled fibers, wrinkles, mis-oriented plies, ply overlaps (butts) or ply underlap/gaps, and even incorrect material can all occur during the lay-up process. Then, during secondary processing, machining defects such as mislocated, damaged, oversize, burnt, or off-axis holes can be introduced—or when joining, over-torqued, undersized, or improperly seated fasteners might be introduced. Moreover, improper management can cause scratches, dents, fiber split, or breaking, as well as surface degradation or delamination [6].

On a microscopic or macroscopic level, composites might fail. In compression buckling, compression failures can happen at both the macro and micro scales. Tension failures can be net section failures of the component or microscopic deterioration of the composite when one or more of the layers in the composite fails in tension of the matrix or failure of the matrix-fiber connection. Many composites are brittle and have minimal reserve strength beyond the initial beginning of failure, whereas others can withstand significant deformations and have a reserve energy-absorbing capacity beyond the initial onset of damage. The wide range of fibers and matrices available, as well as the mixes that can be created with blends, allows for a wide range of characteristics to be built into a composite structure.

Tools for composite manufacturing are significantly less expensive than sheet metal forming tools. This is because composite procedures are one-shot operations (i.e., one mold), whereas sheet metal forming necessitates the use of five to six different tools per component line. These tool cost reductions are particularly important at low production quantities, but at greater volumes—when part prices dominate—this competitiveness is lost. The imperfections seen in natural fibers add to the difficulty. These include regions where the fiber has a growing fault, thinning due to a larger inner hole diameter, and "kink bands" (where the fiber direction is momentarily translated sideways), resulting in local strain concentrations when placed under load. A strain of less than 0.5% is commonly achieved using current techniques to the production of bast fiber composites.

New ways to customize the fiber interface are among the strategies being developed to solve these challenges. Another option is to treat the materials such that the reinforcement dimension is less than the defect level. Because of the high stiffness qualities of cellulose (estimated at 130 GPa), the wood industry is driving the development of nanocellulose across the world, and it may one day create an altogether new type of composite. Major problems connected with nanocellulose dispersion and pre-treatment in its polymer, as well as guaranteeing adherence, may be handsomely rewarded with a flexibly produced material with acceptable environmental properties.

The absence of simulation tools and a general lack of composite material characterization is now a key problem in automobile composite design. Another difficulty is the amount of time it takes to simulate composite structures and components using computers. Current commercial design software composite material models have extremely lengthy solution times. These time frames are typically too long for the first stages of vehicle development, when many alternative choices must be evaluated in just a few months. The automobile sector requires a tenfold decrease in solution times to fully assess composites at this level. Because commercial software developers have yet to address this challenge, some of the most sophisticated research and design centers are developing proprietary techniques that are generally kept private.

When considering the huge production numbers necessary, manufacturing is a problem for composites in the automobile industry. The expense of raw materials is one reason why composites aren't extensively employed in mass production automotive applications, but the primary problem is a lack of adequate manufacturing techniques. Currently, the manufacturing process used is heavily influenced by the desired production rate. A typical truck application may have a volume of 5,000 to 20,000 parts per year, but a typical automobile application may have a volume of 80,000 to 500,000 parts per year, or even more. Tooling costs, scrap output, and cycle time are all other factors to consider [7].

11.3 OPPORTUNITIES FOR RESEARCH AND DEVELOPMENT

Because lightweight materials have the potential to minimize CO_2 emissions and increase resource efficiency in application sectors, they have become an essential element of product design. New lightweight metals and nonmetals are gaining popularity as materials for car and aircraft construction. Boeing recently used carbon-reinforced plastic for the exterior body of the 787 Dreamliner. However, the high cost of these materials is a serious issue for end-user businesses.

Titanium, high strength steel, aluminum, polymers and composites, and magnesium are among the items hailed as the future of the lightweight materials sector. The use of high strength steel in automobiles is rising every day, and it is expected that by 2022, high strength steel will be used in almost two-thirds of all vehicles produced. The strength, corrosion resistance, and formability of lightweight aluminum are enhancing its use in aviation and energy applications. Polymers and composites are gaining popularity as essential materials because they are both strong and attractive.

The aviation sector was the first to use lightweight materials in aircraft design, and the advantages of these materials drew the attention of automakers. Because of the growing demand for alternative energy sources, the newly discovered use in the energy industry is projected to develop significantly. These materials are utilized in the rotor blade components of wind energy equipment. The rise in popularity of renewable energy sources is projected to increase demand for lightweight materials; the lightweight materials industry relies heavily on research and development. Titanium was the focus of several scientific studies following its debut because of its remarkable strength. Titanium's characteristics, on the other hand, are now being discovered to be beneficial in defense and other security applications.

This market's members have already established themselves as material producers. New product development and material innovation have helped companies like SABIC and Cytec Solvay Group establish themselves in the composites industry. Alcoa Inc. and Aleris International, two major aluminum manufacturers, have restructured their product portfolios to accommodate these lightweight materials. The rise in acquisition activity is the most noticeable trend in this sector. Major investment corporations have taken notice of the commercial potential of these materials. Precision Castparts Corp. was bought by Berkshire Hathaway in an all-cash transaction in January 2022. Zhongyang USA LLC and Aleris International concluded a formal agreement in August 2022 for the acquisition of Aleris International.

11.3.1 Industry insights

In 2021, the worldwide lightweight materials market was predicted to be worth USD 113.78 billion, with a CAGR of 8.9% expected during the forecast period (Figure 11.1). Because of growing awareness regarding fuel emissions, many automobile manufacturers are moving to items that lower vehicle weight, which is expected to boost the worldwide market. Over the projected period, rising vehicle demand in North America is likely to boost market growth, and the substantial presence of major auto manufacturers in nations like the United States and Canada will drive this growth.

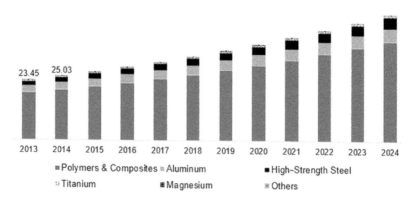

Figure 11.1 World lightweight materials market size [8]. https://www.grandviewresearch.com/industry-analysis/lightweight-materials-market.

Further, the presence of manufacturers of renewable energy equipment is likely to propel the market in this area as well. Europe is also prominent in the lightweight materials market, and the presence of large automobile manufacturers, together with increasing innovation in lightweight materials used in aviation, is likely to drive product demand in this region.

The rising focus on renewable energy will likely increase the use of lightweight materials in the energy sector. Central American countries, for example, are expected to be one of the fastest-growing regions. Growing end-use industries in this area are driving the need for lightweight materials.

Governments are continuously striving to reduce rising levels of pollution caused by auto emissions. Further, rising fuel prices, the adoption of pollution rules, and the financial consequences of not adhering to these requirements are projected to drive demand for lightweight materials in the automotive industry throughout the projection period. To satisfy the industry's safety and emission goals, aluminum, high strength steel, and polymers and composites are commonly employed. Automobile manufacturers are developing multi-material designs, including lightweight materials, to improve vehicle fuel economy. Because of the extra lightweight benefit, the use of polymers and composites in vehicle design is projected to grow.

11.3.2 Product insights

Aluminum, high strength steel, titanium, magnesium, and polymers and composites are among the items that make up the market. Polymers and composites have led this market, and this trend is expected to continue during the projected period. The product's demand is projected to rise as it can cut vehicle weight by 50% and increase fuel economy approximately 35%.

Aluminum has been utilized in a variety of applications since the development of the Hall-Héroult method for producing high-quality aluminum from alumina. Aluminum alloys' age hardening characteristic, along with the strength provided by aluminum, has boosted their use in automotive and aviation applications. Such unique features, as well as their use in a variety of sectors, are expected to fuel market expansion.

In lightweight applications, high strength steel is commonly utilized as a direct replacement for conventional steel. It is widely used in the automotive industry and can reduce vehicle weight up to 25%. High strength steel's recycling value is boosting demand, which is expected to fuel market expansion during the projection period.

Magnesium is mostly utilized in automobiles and mobile gadgets, and its recyclability will drive its market expansion throughout the predicted period. Because of its simple accessibility and plentiful availability, titanium is frequently employed in a number of applications. Due to its exorbitant cost, however, it is only utilized in a few specialized applications where extreme strength and endurance are required.

11.3.3 Application insights

The primary application areas in the lightweight materials market are automotive, aviation, and energy (Figure 11.2). In terms of revenue, the automotive sector led the total lightweight materials market in 2023, accounting for roughly 86% of the market. Aluminum, polymers and composites, and high strength steel are among the most often utilized materials in the automobile sector, and expanding applications in the automotive and aviation sectors are driving the lightweight materials industry. Over the projected period, growing innovation in the aviation sector is likely to boost market demand. Growing application of lightweight materials in the energy application segment is projected to boost market development as well.

The use of lightweight materials has a direct effect on driving dynamics, agility, and fuel economy. Over the projected period, the application of lightweight materials in the aviation industry is likely to enhance demand for lightweight materials. One of the most important renewable energy sources, wind energy, makes use of lightweight materials in the construction of windmills. Transportation and defense are the market's other application segments. Titanium and magnesium are commonly used in these applications because they have the ability to reflect electromagnetic radiation. The market is projected to grow in demand as R&D on novel materials for defense applications increases [9].

Lightweight materials market share, by application, 2016 (%)

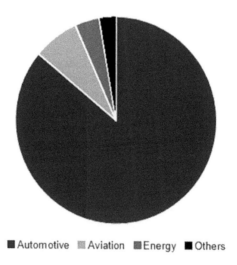

■ Automotive   ▨ Aviation   ▩ Energy   ■ Others

Figure 11.2 Future applications of lightweight material [8]. https://www.grandviewresearch. com/industry-analysis/lightweight-materials-market.

11.3.4 Lightweight materials market share insights

The lightweight materials industry is competitive because of prominent firms such as Cytec Solvay Group, SABIC Industries, Alcoa Inc., and Toray Industries, Inc. The market is fragmented, and mergers and acquisitions are more and more common. Companies are investing in improving product quality so as to enhance market demand. The amount of polymers and composites producers is projected to rise throughout the forecast period because of expanding use in the automotive and aviation industries. The fast-growing wind energy equipment sector has also drawn a slew of new entrants, who have begun to diversify their product portfolios to include certain items utilized in blade manufacture [10].

11.4 SCOPE FOR ADVANCED LIGHTWEIGHT MATERIALS

Lightweight materials are used in a variety of sectors, including automotive, aircraft, wind energy, electrical, sports, household purposes, civil construction, and medical and chemical industries. They offer a lot of promise for use in constructions that can be subjected to compression stresses. Advanced lightweight materials offer appealing characteristics such as high compressive strength, flexibility in constructing thick composite shells, low weight, low density, and corrosion resistance. Because of their superior characteristics, lightweight materials are used to build various elements of automobiles and airplanes. Furniture, windows, doors, matting, civil building, and other household items are also made of lightweight materials.

Although lightweight building materials have a lower density than water, they have a better strength-to-weight ratio than aluminum or steel. Modern materials such as composite materials for lightweight construction are utilized in various industries to decrease manufacturing costs, enhance product quality and usefulness, and boost durability. Lightweight building materials also aid in the optimization of production processes and material usage for optimum efficiency.

11.4.1 Advanced lightweight materials market size and segment forecast, 2019–2026

The worldwide lightweight materials market was worth USD 131.0 billion in 2018, and it is expected to expand at a CAGR of 7.9% over the next five years (Figure 11.3). Magnesium, carbon fiber, aluminum/Al composites, titanium, glass fiber, and high strength steel are just a few examples of lightweight materials. These materials in automotive and aerospace reduce weight, improving fuel efficiency and extending the range of hybrids, plug-in hybrids, and all-electric vehicles.

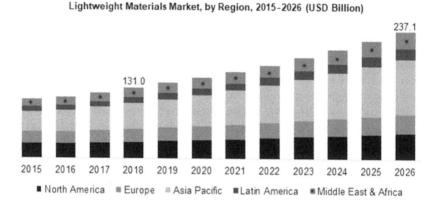

Figure 11.3 World market of lightweight materials [11]. https://www.polarismarketresearch.com/industry-analysis/lightweight-materials-market.

The use of lightweight structural materials allows for sophisticated emission control systems, safety devices, and integrated electrical systems without increasing the vehicle's total weight or decreasing its efficiency. Titanium is a high-temperature metal that may save weight up to 55% in powertrain systems. It has a high strength-to-weight ratio as well as the capacity to endure extreme temperatures. In strength-limited designs like pillars and door rings, Advanced High Strength Steel can reduce component weight up to 25%. It has excellent strength, stiffness, and formability, as well as good corrosion resistance and low cost.

The growing automotive sector and increasing vehicle modernization are two main factors driving the lightweight materials market. The expansion of the lightweight materials sector is aided by government rules concerning efficiency and road and vehicular safety. The use of lightweight materials is increasing in windmill projects, and the penetration of lightweight components is increasing.

The entire market has grown due to rising demand for aircraft modules and the growing demand for electric cars. The market is expected to rise due to the further development of autonomous cars and self-driving vehicles, as well as increased demand from the aviation sector and investment in R&D. Factors such as rising demand from growing economies such as China, Japan, and India, as well as technology developments, are likely to create even more growth possibilities.

11.4.2 Product analysis

The total market is divided into three categories based on product: metal alloys, composites, and polymers. High strength steel, aluminum, titanium, and magnesium are all included in the metal alloys section. Carbon fiber-reinforced polymers (CFRP) and glass fiber-reinforced polymers (GFRP) are separated in the composites section (GFRP). Polycarbonate and polypropylene are separated from the polymers section. The polymers and composites sectors led the entire market, and this trend is expected increase since they can lower vehicle weight by more than 45%, resulting in a 30% increase in fuel economy.

In many applications that need lightweight material, high strength steel is utilized as an alternative for standard steel. This is owing to the material's high tensile strength. It is widely used in the automobile sector and has the potential to reduce overall vehicle weight by about 25%.

Aluminum is another material utilized in a variety of applications that call for lightweight materials. Age hardening is a characteristic of aluminum alloys, and they are equally strong. These characteristics have boosted demand in the aviation and automobile industries. Aluminum is a common material in aviation used to make structural components. Such features of this material, as well as its employment in a variety of end-use industries, are anticipated to lead to segment development.

Magnesium alloys may reduce weight by up to 25% while also improving fuel economy by up to 1%. They've been utilized in mobile electronic devices, for example. Titanium is a long-lasting material used in a variety of high-strength applications. It is mostly utilized in specialized applications and is more expensive than other materials on the market.

11.4.3 End-use analysis

Aerospace, automotive, marine, construction, and other industries are the end-user segments in the lightweight materials business. Because of vehicle modernization and the rising penetration of electric cars, the auto industry led the worldwide lightweight materials market in 2021. Manufacturers are using lighter components in automobiles to enhance mileage due to increased awareness about gas emissions. Aluminum, polymers and composites, and high strength steel are all important components in this application. Driving agility, dynamics, and fuel economy are all affected by the usage of these materials. Furthermore, OEMs are being compelled to use lighter parts to produce automobiles that produce less carbon dioxide emissions.

11.4.4 Regional analysis

The market for lightweight materials is divided into three categories: type, application, and geography. Metal alloys, composites, and polymers are the three types of materials used in the market. It is divided into automotive, aerospace, wind, marine, and other applications according to its use. North America, Europe, Asia-Pacific, and Latin America are the regions studied.

In 2021, Asia-Pacific held the biggest share of the global lightweight materials market. Growing disposable income and rising living standards have boosted demand for automobiles in the Asia-Pacific area, supporting market development. During the forecast period, the adoption of rigorous government laws governing vehicular safety and the development of autonomous cars are likely to create growth possibilities in the Asia-Pacific market.

High strength steel is in high demand in North America since it has similar characteristics to conventional steel. Composites and polymers, on the other hand, are anticipated to develop rapidly in the region due to widespread adoption in the energy and automotive industries. In specific applications, titanium and magnesium are utilized. Aluminum consumption is increasing due to its use in the aviation and automotive industries [12].

11.4.5 The future scope of aluminum alloys

Over the next ten years, demand for aluminum is expected to double. There will be an 80-million-tonne worldwide demand by 2025. As a result, the aircraft industry is increasingly turning to recycled alloys to meet the growing demand. There is also a push for innovation in terms of the materials utilized and the construction of airplanes. Aluminum-lithium alloys, for example, have been created for the aerospace sector to reduce aircraft weight and therefore enhance performance. Al-Lithium alloys, because of their low density, high specific modulus, and outstanding fatigue and cryogenic toughness qualities, have become indispensable in aircraft construction. As emerging countries begin to invest more in the aerospace sector in the next few years, there will be more innovation in aluminum alloys [13].

The supply of aluminum, as well as other alloying elements such as zinc, copper, magnesium, manganese, and tin, is critical to the global aluminum alloys market. Fluctuations in the price of aluminum regarding its application in a variety of end-user sectors may limit the growth of the Global Aluminum Alloys Market in the forecast years. Due to the competitive market, price volatility of other alloying elements such as copper, zinc, and magnesium may also be an impediment to the industry's expansion [14].

11.4.6 The future scope of high-speed steel

During the assessment period of 2021 to 2031, the worldwide high-speed steel market is expected to grow at a CAGR of close to 7%, reaching a valuation of US$4.5 billion in 2031. Over the next 10 years, M grade high-speed steel will have a market share of more than 40%. Great-speed steel has a wide range of applications and specific product lines and is in high demand in a variety of end-use sectors. With comprehensive application and diverse use cases, the steel market is a driving force for high-speed steel.

High-speed steel has shown absolute growth at a CAGR of over 4% from 2016 to 2020 over the last half-decade. This expansion is fueled by new building projects, as well as end-use industries and a variety of applications, such as cutting tools, metalworking, and milling. In the second and third quarters of FY2020, the COVID-19 epidemic reduced demand for high-speed steel, and lockdowns and shutdowns in major end-use industries such as construction, manufacturing, and automotive generated a domino effect on high-speed steel requirements around the globe. Overall demand, however, increased by 2.3% in the fourth quarter of 2020. Taking into account all the results for the market throughout the forecast period, demand is predicted to increase steadily in the short term (2020–2025), while modest growth is predicted in the medium term. Overall, the market is expected to grow at a CAGR of over 7% through 2031 [15].

11.4.7 Advanced composite materials

The advanced composites industry is projected to grow at a healthy rate as a whole. The advanced composite market was valued at $16.3 billion in 2020, with a predicted CAGR of 7.2% from 2020 to 2024. In 2014, North America owned the biggest share of the market by volume. The expanding list of applications in the defense sector as well as the aerospace industry, which will be my emphasis throughout this chapter, will be mainly responsible for the region's growth. Compared to previous years, the aerospace market for composite materials is expected to develop at a faster rate during the next decade (Figure 11.4).

Next-generation airplanes, such as the Boeing 787, are demonstrating these new uses that the composites industry had previously overlooked. By employing these sophisticated materials, the Boeing 787 saves almost 20%

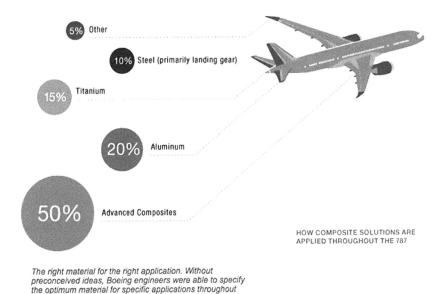

5% Other

10% Steel (primarily landing gear)

15% Titanium

20% Aluminum

50% Advanced Composites

HOW COMPOSITE SOLUTIONS ARE
APPLIED THROUGHOUT THE 787

The right material for the right application. Without preconceived ideas, Boeing engineers were able to specify the optimum material for specific applications throughout the airframe.

Figure 11.4 Usages of composite materials in aerospace industry.

of the aircraft's frame weight when compared to a standard aluminum frame. Because of the weight savings and the lower fuel consumption of this new composite-driven airplane, more travel is feasible. Composite materials not only save fuel and resist corrosion, but they may also provide lightning protection, via a composite coating developed by a group of researchers.

This growth in the use and applications of sophisticated composite materials will also be seen in the defense industry. Although many applications will be in the form of aircraft and aircraft components for light weighing and strength, as previously noted, there are other novel uses that will benefit the military industry. From sophisticated bullet-stopping personal protective vests to a new range of vehicle armor for both personal protection and military uses, there's something for everyone. Morgan Advanced Materials (MAM), for example, utilizes these composites to create light armor such as helmets, ballistic protection jackets, and vests. MAM claims its armor is 50% lighter than similar steel armor, while still providing protection against threats such as mine explosions, flaming fluids, and fragmentation [16].

11.5 RECOMMENDATIONS AND CONCLUSION

Modern vehicles require advanced materials to improve fuel efficiency while preserving safety and performance. Lightweight materials have a lot of potential for improving vehicle economy since they require less energy to accelerate a lighter item than a heavy one. A 10% reduction in vehicle weight might result in a 6%–8% increase in fuel economy. By replacing cast iron and conventional steel components with lightweight materials including high strength steel, Mg alloys, Al alloys, carbon fiber, and polymer composites, a vehicle's body and chassis may be reduced by up to 50%, lowering its fuel consumption.

By 2030, using lightweight components and high-efficiency engines made possible by new materials in a fourth of the US fleet may save more than 5 billion gallons of gasoline each year. In the near term, replacing heavy steel components with high strength steel, aluminum, or glass fiber-reinforced polymer composites can reduce component weight by 10% to 60%. Although lightweight construction aids in weight reduction, the quick success of simple material substitution conceals the reality that it is far from easy. Magnesium is significant in this context since it is the best material for castings.

For years, lightweight building has been a hot issue; although lightweight architecture generally achieves more weight savings than a simple material replacement, in reality, designers attempt to employ a variety of materials in lightweight construction. However, a simple material swap is rarely a satisfactory solution since a new material typically necessitates a different design (for example, greater installation space or different radii) and/or manufacturing (for example, forming, original forming, or joining technique).

In terms of industry, lightweight construction must not only be the optimum technical compromise between weight and other technological criteria, but it must also be cost-effective. The majority of mechanical engineering products are constructed of metal for economic reasons. Steel, aluminum, and magnesium are the most essential materials. Their different material qualities affect not just quality indicators but also the processability of semi-finished and completed goods, as well as the costs connected with them.

REFERENCES

1. Tasuns composite technology, https://www.china-composites.net/news/advantages-disadvantages-of-carbon-fiber-1585756.html. 2016.
2. ASAP Aviation Procurement Blog, https://www.asap-aviationprocurement.com/blog/pros-and-cons-of-composite-materials-on-aircraft/. 2019.
3. Global electronic service, https://gesrepair.com/what-are-the-benefits-and-challenges-of-lightweight-manufacturing/. 2021.
4. James R. Fekete. "Manufacturing challenges in stamping and fabrication of components from advanced High Strength Steel," *International Symposium on Niobium Microalloyed Sheet Steel for Automotive Application*, 2006.
5. Green spec, https://www.greenspec.co.uk/building-design/aluminium-productionenvironmental-impact/. 2021.
6. Greenhalgh, E. "Failure analysis and fractography of polymer composites," *Woodhead Publishing Series in Composites Science and Engineering* 2009, 356–440. https://doi.org/10.1533/9781845696818.356
7. Haruna V. N., Abdulrahman A. S., Zubairu P. T., Isezuo L. O., Abdulrahman M. A., Onuoha D. C. "Prospects and Challenges of Composites In a Developing Country," *Journal of Engineering and Applied Sciences* 9(7):1070–1075.
8. Lightweight Materials Market Size Report Analysis by Product (Aluminum, Polymers & Composites), by Application (Automotive, Aviation, Energy), by Region, and Segment Forecasts, 2017–2024.
9. Grand View research, https://www.grandviewresearch.com/research-insights/lightweight-materials-material-innovation-future. 2016.
10. Grand View research, https://www.grandviewresearch.com/industry-analysis/lightweight-materials-market. 2019.
11. Lightweight Materials Market Share, Size & Trends Analysis Report by Product (Metal Alloys, Composites, Polymers); by Application (Aerospace, Automotive, Construction, Energy, Aviation); by Region: Market Size and Segment Forecast, 2019–2026.
12. Polaris Market Research, https://www.polarismarketresearch.com/industry-analysis/lightweight-materials-market. 2021.
13. Matmatch, https://matmatch.com/blog/aluminium-alloys-in-aerospace-industry/. 2021.
14. Energy and Environment from open pr. https://www.openpr.com/news/1769916/future-scope-of-aluminium-alloysmarket-2019-2023-estimated-to-grow-cagr-rapidly-analysis-by-global-top-key-company-s-alcoa-rio-tinto-alcan-kaiser-aluminum-aleris-rusal-constellium-ami-metals.html. 06-10-2019 02:32 PM

CET | Energy & Environment Press release from: Business Industry Reports / PR Agency: Business Industry Reports.

15. Fact.MR, https://www.factmr.com/report/883/high-speed-steel-market. March 2021.

16. The Future of Advanced Composite Materials October 12, 2015, Garret Carden, Market Analyst and Development Intern at Mar-Bal, Inc. and an Industrial Engineering student (Senior) at West Virginia University https://www.mar-bal.com/language/en/future-advanced-composite-materials/

Index

For Product Safety Concerns and Information please contact our EU
representative GPSR@taylorandfrancis.com
Taylor & Francis Verlag GmbH, Kaufingerstraße 24, 80331 München, Germany

www.ingramcontent.com/pod-product-compliance
Ingram Content Group UK Ltd.
Pitfield, Milton Keynes, MK11 3LW, UK
UKHW021118180425
457613UK00005B/145